Materials for Technology Students

Materials for Technology Students

Vernon John

M.Sc., C.Eng., M.I.M.M., A.I.M.

Senior Lecturer in Engineering,
The Polytechnic of Central London

First edition 1975
Reprinted 1978

Published by
THE MACMILLAN PRESS
London and Basingstoke
Associated companies in Delhi Dublin
Hong Kong Johannesburg Lagos Melbourne
New York Singapore and Tokyo

ISBN 0 333 17641 3

Printed and bound in Great Britain by
REDWOOD BURN LIMITED
Trowbridge & Esher

Contents

Preface

In recent years there have been a number of significant changes in the approach to education for engineering. One of the most important of these changes has been the recognition that the science of materials must form an integral part of any engineering course. In the past, the intensive study of materials normally commenced in Higher National or degree courses. I am very pleased to see that the subject is now introduced in its rightful place as one of the basic foundation studies in the O.N.D. in Technology course and also at G.C.E. Advanced level in engineering science courses. Unfortunately, the O.N.D. in Technology course is only to be a transient phenomenon and will shortly be superseded by the new Technician Diploma and Certificate courses. However, materials science studies will feature prominently in the new courses and technician students should find the style and content of this book suitable for them.

In this book I have used a largely descriptive approach to the very broad subject of materials which should be suitable for students with a background of G.C.E. Ordinary-level science. Some knowledge of differential and integral calculus is needed for the work on elementary stress analysis, but by the time this stage of the materials course is reached the student should be conversant with these techniques from his complementary studies in mathematics.

A number of test questions have been placed at the ends of chapters 5, 6 and 7, the stress analysis section. I have not included revision questions for other chapters because, in very many instances, colleges choose to examine at O.N.D. level by means of multiple-choice objective tests.

I wish to record my gratitude for the help and advice received from colleagues and family during the preparation of this book, in particular Clive Beesley and Ed Drabble of P.C.L. for reading the manuscript and suggesting improvements. I am greatly indebted to my wife for reducing my untidy handwriting to a clean typescript, to my elder son Roger for invaluable assistance in proof reading, and to my younger son Derek for not interrupting too often while the work was in progress.

Vernon John

List of Symbols

Quantity	Symbol
Atomic number	Z
Atomic mass number (atomic weight)	M
Avogadro's number	$N_A = 6.023 \times 10^{23}$ (molecules per mole)
Planck's constant	$h = 6.625 \times 10^{-34}$ Js
Temperature	T or θ
Thermal conductivity	k
Universal gas constant	$R_0 = 8.314$ kJ/kmol K
Frequency	ν (Greek nu)
Direct stress	σ (Greek sigma)
Direct strain	ϵ (Greek epsilon)
Shear stress	τ (Greek tau)
Shear strain	γ (Greek gamma)
Young's modulus of elasticity	E
Modulus of rigidity	G
Bulk modulus of elasticity	K
Poisson's ratio	ν
Force	F
Bending moment (moment of force)	M
Second moment of area	I
Polar second moment of area	J
Torque	T
Power	P
Elastic strain energy	U
Time	t
Glass transition temperature	T_g
Surface tension	γ
Velocity of light	$c = 3 \times 10^8$ m/s

Units

The units used through this book conform to the SI system. The principal units
that are quoted in the text are given below. (Since we are in the change-over
period, the Imperial equivalents of some SI units are listed.) Preferred SI units are
printed in bold type.

Quantity	*Unit*	*Symbol*
Mass	**kilogram**	**kg** (1 kg = 2.205 lb)
	gram	g
	tonne	t (Mg) (1 t = 1000 kg = 0.984 ton)
Length	**metre**	**m** (1 m = 39.37 in.)
	millimetre	mm
Time	**second**	**s**
	minute	min
	hour	h
Velocity	**metre per second**	**m/s**
Temperature	degree Kelvin	K
	degree Celsius	°C
Amount of substance	**mole**	**mol**
	kilomole	kmol
Plane angle	**radian**	**rad** (1 rad = $180°/\pi$)
	degree	°
Area	**square metre**	**m²**
	square millimetre	mm²
Volume	**cubic metre**	**m³** (1 m³ = 35.315 ft³)
	cubic millimetre	mm³
Density	**kilogram per cubic metre**	**kg/m³** (1 kg/m³ = 1000 g/cm³ = 0.062 lb/ft³)
Force	**newton**	**N** (1 N = 0.225 lbf)
	kilonewton	kN
	meganewton	MN

Quantity	*Unit*	*Symbol*
Moment of force (Torque)	**newton metre** kilonewton metre	**Nm** (1 Nm = 0.738 lbf ft) kNm
Stress (pressure)	**newton per square metre**	**N/m²** (1 N/m² = 0.000145 lbf/in²)
	meganewton per square metre	MN/m² (1 MN/m² = 0.102 kgf/mm² = 0.0648 tonf/in²)
	giganewton per square metre	GN/m²
	bar	bar or b (1 bar = 10⁵ N/m²)
Surface tension	newton per metre	N/m
Energy	**joule**	**J** (Nm)
	electron volt	eV (1 eV = 1.602 × 10⁻¹⁹ J)
Calorific value (mass)	megajoule per kilogram	MJ/kg (1 MJ/kg = 429.5 Btu/lb)
Calorific value (volume)	megajoule per cubic metre	MJ/m³ (1 MJ/m³ = 26.81 Btu/ft³)
Thermal conductivity	watt per metre per degree Kelvin	W/m K
Power	watt	W (J/s)
	kilowatt	kW (1 kW = 1.342 hp)
Second moment of area	metre to the fourth	m⁴
	millimetre to the fourth	mm⁴
Electric current	**ampere**	**A**
Voltage	volt	V
Quantity of electricity	coulomb	C (As)
Electrical resistance	ohm	Ω (V/A)
Electrical resistivity	ohm metre	Ωm
Magnetic flux	weber	Wb (Vs)
Magnetic flux density	tesla	T (Wb/m²) (1 T = 10⁴ gauss)
Frequency	hertz	Hz (s⁻¹)

1

Atomic Structure and Bonding

1.1 Introduction

Materials are the very foundations of technology, since without suitable materials nothing can be constructed. A portion of a dictionary definition of an engineer is: 'one who designs or carries out the construction of works of public utility; one who constructs engines.' It will be apparent that no matter how brilliant a new piece of engineering design may be, it is of no avail if there are no suitable materials available for converting the design into 'hardware'.

At the dawn of history man could only use those materials naturally available to him — stones, mud and clay, fallen tree-trunks and animal and vegetable fibres. With the development of tools, stone and timber could be dressed into the desired shape. Then methods of producing iron and the alloy bronze were discovered and the art of forging these metals into various shapes was evolved. One early constructional material was sun-dried mud brick and it was soon realised that a brick with better properties could be obtained by mixing straw with the mud or clay before shaping and drying. This same principle is used today for fibre-reinforced plastics.

Many metals including copper, gold, iron, lead, silver and tin were produced and used by early civilised peoples. It was discovered that the properties of a metal could be altered by alloying with other metals and a large number of useful alloys was developed by these empirical means. About 1000 years ago the first steels were produced by heating iron bars in charcoal, allowing carbon to infuse into the iron. However, development was slow; new designs and new forms of construction could not proceed in advance of materials development. As an example, the major constructional materials of earlier centuries — stone and brick — while strong in compression are weak in tension. The arch is a design in which a gap may be spanned by a masonry structure, because in an arch the individual bricks or stones are stressed in compression, see figure 1.1. There is, however, a limit to the span possible in a masonry arch. Brunel's brick bridge over the River Thames at Maidenhead possesses two flat arches, each span being 39 m; these are the largest brick arches that have been constructed. The largest masonry arch is at Plauen in East Germany, with a span of almost 90 m. The development of bridge design could not progress from the arch until the latter part of the nineteenth century when iron and steel materials, whose strengths in tension are comparable to those in compression, became widely available.

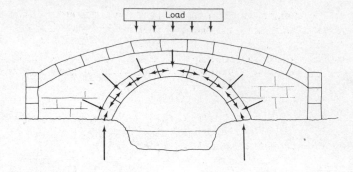

FIGURE 1.1   *Masonry arch*

During the last 100 years or so, enormous developments have occurred in the materials field. Metals such as aluminium, magnesium and nickel which were curiosities at one time are now commonplace; many new alloys have been produced; a complete new industry — the plastics industry — has been created; industrial ceramics have been developed. While early development was largely empirical there has been intensive study of the science of materials during this century, and the present-day engineering materials are soundly based on scientific theory and practice.

Since there are many thousands of different materials available today it is impossible for the technologist to have detailed knowledge of them all, but it is essential that he has a good grasp of the fundamental principles that control the properties of materials. He will also have to consider many factors other than the mechanical, electrical and thermal properties of a material before he can select the most suitable material for some specific task. These include possible alteration of the properties of a material with time in service, corrosion and chemical change, form and availability of raw material supply, restrictions on processing and forming to shape, and, of course, the cost of material and processing.

There is one thing that all materials have in common, and that is that they are all composed of atoms. The atoms themselves may differ from one another, but the atoms of all 103 known elements are composed of the same three basic particles — the *proton*, the *neutron* and the *electron*. It is the way in which these particles are assembled into atoms, and how the atoms are bonded to one another that largely control the final properties of the bulk engineering materials we use every day.

1.2 Elementary Particles

Atoms are made up of three types of elementary particle. These are the proton, the neutron and the electron. The proton is a particle that has a positive electrical charge of 1.602×10^{-19} C and a mass of 1.672×10^{-24} g. The neutron has an almost identical mass (1.675×10^{-24} g) but has no electrical charge. The third particle, the electron, has an electrical charge that is equal and opposite to that of the proton but has a mass of only $1/1838$ of that of the proton, namely 9.107×10^{-28} g.

The atom as a whole is electrically neutral and therefore contains protons and electrons in equal number. The number of neutrons in the atom is often the same as, or greater than, the number of protons.

1.3 Atomic Number and Atomic Mass Number

Each of the 103 different elements is characterised by a different number of protons in the nucleus. The number of protons contained in the nucleus is termed the *atomic number*, Z. In an electrically neutral atom there must, therefore, be Z electrons.

The nucleus is normally composed of protons and neutrons, and both of these particle types make up the mass of the atom. The *atomic mass number* or *atomic weight*, M, indicates the total number of protons and neutrons in the nucleus. For the simplest atom, hydrogen, $Z = 1$, $M = 1$, so that the nucleus is just one proton. For oxygen, $Z = 8$, $M = 16$, so that the nucleus contains eight protons and eight neutrons.

The quoted atomic mass numbers for atoms are not all integers. (The atomic weight of hydrogen is 1.008.) One of the reasons for this is the existence of *isotopes*. Isotopes of elements possess the characteristic atomic number, but have differing mass numbers. In the case of hydrogen the following isotopes are known

$$Z = 1 \qquad M = 1 \qquad \text{normal hydrogen}$$
$$Z = 1 \qquad M = 2 \qquad \text{deuterium}$$
$$Z = 1 \qquad M = 3 \qquad \text{tritium}$$

1.4 The Gram Atom and Avogadro's Number

The gram atom of an element is the amount of the element whose mass in grams is equal to its atomic mass number.

Similarly a gram molecule, or *mole*, is the amount of compound whose mass in grams equals its molecular mass number (molecular weight), the molecular mass number being the sum of the atomic mass numbers of the constituent atoms within the molecule.

The term mole (abbreviation mol) is used as an alternative name for both the gram atom of a monatomic element and the gram molecule of a molecular substance. A larger unit may also be used in connection with the molar mass of any substance. This larger unit is the kilogram molecule, or *kilomole* (abbreviation kmol).

One mole of any substance will always contain the same number of molecules and the number of molecules contained in a mole is given by the universal constant, Avogadro's number, N_A. The numerical value of Avogadro's number is 6.023×10^{23}/mol (or 6.023×10^{26}/kmol).

1.5 The Structure of the Atom

In 1911 Rutherford performed a series of experiments in which he bombarded thin gold foil with α-particles (helium nuclei, which are positively charged particles with an atomic weight of 4). Very many α-particles passed through the foil without being deflected, but some particles were deflected. The inference was that much of the atom was void space allowing most of the α-particles to pass straight through, but with some particles deflected because they passed close to, or collided with, a central nucleus of very small dimensions. The effective diameter of a gold atom is approximately 3×10^{-10} m and the diameter of the nucleus is about 1/10 000 of this size.

One of the first postulations made was that the electrons are in orbit around the central nucleus in much the same way as planets are in orbit around the sun. On the basis of classical mechanics this picture cannot be valid. The electron is a charged particle and in following a circular or elliptical path will be subject to angular acceleration. But from electromagnetic theory an electrical charge will radiate energy when accelerated; (it is the acceleration of charges in an aerial that is responsible for the transmission of energy as radio waves). This would not allow stability in atoms; if orbiting electrons were continually losing energy, the orbits would not remain constant; the electrons would follow spiral paths and eventually collapse into the nucleus.

Electrons, therefore, cannot simply be regarded as corpuscular particles. They exhibit some of the characteristics of waves — an electron beam can be diffracted — and the motion of electrons can be described, not by classical mechanics, but by quantum mechanics and wave mechanics.

In 1913 Bohr made the suggestion that an electron always moved in a closed orbit and as long as it did so there would be no emission or absorption of energy. More than one stable orbit or energy level is possible for any electron. According to Bohr's theory, within any stable orbit, or energy level, the angular momentum of the electron must be an integral multiple of $h/2\pi$, where h is Planck's constant. Bohr also postulated that an electron may jump from one stable orbit to another and in so doing will emit or absorb energy at a definite frequency.

If atoms of an element are excited in some way, for example in a gas-discharge tube or in an electric arc, radiation is emitted and this may be analysed with the aid of a spectroscope. Each element emits its own characteristic spectrum, but the spectra are line spectra, rather than a continuous band form. Spectral analysis in the late nineteenth and early twentieth centuries showed that for any element there were four series of spectral lines.

In 1900 Planck had postulated that energy is not emitted or absorbed in a continuous manner, but rather in discrete packets termed *quanta* with one quantum of energy being $h\nu$, where h = Planck's universal constant and ν = frequency of the radiation.

According to Bohr, when an atom is excited by the input of energy an electron moves to an orbit of greater radius and hence greater energy. Subsequently when the excited electron jumps back from some high energy level E_1 to some lower energy level E_2 the energy released will give rise to a spectral line of frequency ν, given by Planck's quantum relationship

$$E_1 - E_2 = h\nu$$

1.6 The Periodic Table

As stated in section 1.3, atoms of each of the 103 different chemical elements possess Z electrons, where Z is the atomic number of the element. The electrons are arranged around the nucleus in a series of shells or electron groups, but the capacity of each electron shell is not the same. The first, or K shell, is filled when it contains 2 electrons. The second shell, the L shell, has a complement of 8 electrons and the third shell, the M shell, may hold up to 18 electrons. Sub-shells or electron sub-groups occur in all electron shells other than the K shell. These sub-groups are designated by the letters s, p, d and f. The maximum number of electrons that can be placed into the various sub-groups are as follows: s sub-group 2 electrons, p sub-group 6 electrons, d sub-group 10 electrons, f sub-group 14 electrons. This information is summarised in table 1.1.

TABLE 1.1
Number of electrons possible for shells and sub-shells

Shell	Sub-shells (numbers indicate maximum electron capacity)				Total number of electrons in completely filled shell
	s	p	d	f	
1st or K	2	–	–	–	2
2nd or L	2	6	–	–	8
3rd or M	2	6	10	–	18
4th or N	2	6	10	14	32

The theoretical electron capacity of the fifth, sixth and seventh electron shells would be 50, 72 and 98 respectively, but in practice no electron shell contains more than 32 electrons. The higher atomic number elements possess electrons in up to seven shells but the outer shells are incompletely filled.

The electronic configuration of the first 38 elements, in ascending order of atomic number, are given in table 1.2.

As the atomic number increases, at least up to argon ($Z = 18$), each additional electron takes up a place in the appropriate shell, filling first the 1s, then 2s and 2p, followed by 3s and 3p. When the 3p sub-shell is completed, the nineteenth electron (for potassium) goes not into the 3d sub-shell, but into the 4s sub-group, since this state is at a slightly lower energy level than 4p. This means that after element number 20 (calcium), when the 4s state is full, additional electrons go into the 3d state. In passing from elements 21 to 30 (scandium to zinc) the 3d sub-shell is completely filled and with element 31 (gallium) we find that the 4p sub-shell is started.

For all sub-groups above 4s, the energy levels are fairly close together and the general order in which the shells fill up with electrons is

4s, 3d, 4p, 5s, 4d, 5p, 6s, 4f, 5d, 6p, 7s, 6d, and 5f

TABLE 1.2

Electronic configuration of the elements

Element	Symbol	Atomic number Z	Atomic weight M	Number of electrons in each shell and sub-shell																		
				K	L		M			N				O				P				Q
				1s	2s	2p	3s	3p	3d	4s	4p	4d	4f	5s	5p	5d	5f	6s	6p	6d	6f	7s
hydrogen	H	1	1.008	1																		
helium	He	2	4.003	2																		
lithium	Li	3	6.940	2	1																	
beryllium	Be	4	9.013	2	2																	
boron	B	5	10.82	2	2	1																
carbon	C	6	12.011	2	2	2																
nitrogen	N	7	14.008	2	2	3																
oxygen	O	8	16.000	2	2	4																
fluorine	F	9	19.000	2	2	5																
neon	Ne	10	20.183	2	2	6																
sodium	Na	11	22.991	2	2	6	1															
magnesium	Mg	12	24.32	2	2	6	2															
aluminium	Al	13	26.98	2	2	6	2	1														
silicon	Si	14	28.09	2	2	6	2	2														
phosphorus	P	15	30.975	2	2	6	2	3														
sulphur	S	16	32.066	2	2	6	2	4														
chlorine	Cl	17	35.457	2	2	6	2	5														
argon	Ar	18	39.944	2	2	6	2	6														

				1s	2s	2p	3s	3p	3d	4s	4p	5s
potassium	K	19	39.10	2	2	6	2	6		1		
calcium	Ca	20	40.08	2	2	6	2	6		2		
scandium	Sc	21	44.96	2	2	6	2	6	1	2		
titanium	Ti	22	47.90	2	2	6	2	6	2	2		
vanadiu	V	23	50.95	2	2	6	2	6	3	2		
chromium	Cr	24	52.01	2	2	6	2	6	5	1		
manganese	Mn	25	54.94	2	2	6	2	6	5	2		
iron	Fe	26	55.85	2	2	6	2	6	6	2		
cobalt	Co	27	58.94	2	2	6	2	6	7	2		
nickel	Ni	28	58.69	2	2	6	2	6	8	2		
copper	Cu	29	63.54	2	2	6	2	6	10	1		
zinc	Zn	30	65.38	2	2	6	2	6	10	2		
gallium	Ga	31	69.72	All K, L and M shells are completely filled						2	1	
germanium	Ge	32	72.60							2	2	
arsenic	As	33	74.91							2	3	
selenium	Se	34	78.96							2	4	
bromine	Br	35	79.916							2	5	
krypton	Kr	36	83.80							2	6	
rubidium	Rb	37	85.48	2	2	6	2	6		2	6	1
strontium	Sr	38	87.63	2	2	6	2	6		2	6	2

The number of electrons in the outermost shell has a very important bearing on the properties of the element. Some elements, those with atomic numbers of 2, 10, 18, 36, 54 and 86, are extremely stable and inert chemically. These, the inert gases or group 0 elements, are characterised by having all their electron sub-groups completely filled with electrons. Apart from helium, which has only 2 electrons, these elements all have a complement of 8 electrons in their outermost principal shell.

All other elements, which have at least one incompletely filled shell or sub-shell, will take part in chemical reactions. It is the outermost-shell electrons — the valency electrons — that are responsible for the chemical reactivity of elements, and elements in combination attempt to achieve an outer-shell configuration containing 8 electrons.

It was noticed more than a hundred years ago, long before deductions had been made about the inner structure of the atom, that there existed certain groups of atoms showing some similarity in properties. The alkali metals, lithium, sodium and potassium formed one such group, and the halogens, fluorine, chlorine, bromine and iodine, formed another. When the various elements known at that time were placed in order of ascending atomic weight, a certain periodicity of elements with specific properties, such as the alkali metals and the halogens, was observed. In 1870 Mendeleev arranged the elements in a table in a manner which highlighted this effect. This was the forerunner of the periodic table of elements as it is known today. With the discovery of the electron and the build-up of electrons in shells and sub-shells as outlined above, came a fuller understanding of the properties of the elements.

Table 1.3 gives the full periodic table of the elements. The figures in front of each row of elements, for example, 3p in front of elements 13 to 18, indicate the particular sub-shell being filled with electrons. The electrons in the outermost shell are the valency electrons. For the vertical columns IA, IIA, IIIB, IVB, VB, VIB, VIIB and 0, within each column the elements possess the same number of outer-shell electrons, and hence, possess similar properties, but considerable property differences occur between the elements of one column and those of another.

In the transition series, the first of which is the horizontal row containing the elements 21 to 30, the sub-shell being filled is not in the outer shell. This means that all the elements in the row have somewhat similar properties. However, because the 4s, 4p, and 3d energy levels are close to one another, there is a tendency for electrons to move from one sub-shell to another. This causes the elements to possess variable valency characteristics. For example, iron has a valency of 3 in the ferric state and a valency of 2 in the ferrous state.

In the case of the two long series, the lanthanides, elements 58 to 71, and actinides, elements 90 to 103, the sub-shell being filled is well in towards the core of the atom and so the chemical differences between each member of these two groups is very small.

1.7 Valency

The electronic arrangement found in the inert gases of group 0 is exceptionally stable and in chemical reactions atoms of other elements tend to form the electronic configuration of 8 electrons in the outermost shell (2 electrons, the helium

TABLE 1.3

PERIODIC TABLE OF THE ELEMENTS

	IA	IIA	IIIA	IVA	VA	VIA	VIIA	VIII			IB	IIB	IIIB	IVB	VB	VIB	VIIB	O
1s	1 H																	2 He
2s / 2p	3 Li	4 Be											5 B	6 C	7 N	8 O	9 F	10 Ne
3s / 3p	11 Na	12 Mg											13 Al	14 Si	15 P	16 S	17 Cl	18 Ar
4s / 3d / 4p	19 K	20 Ca	21 Sc	22 Ti	23 V	24 Cr	25 Mn	26 Fe	27 Co	28 Ni	29 Cu	30 Zn	31 Ga	32 Ge	33 As	34 Se	35 Br	36 Kr
5s / 4d / 5p	37 Rb	38 Sr	39 Y	40 Zr	41 Nb	42 Mo	43 Tc	44 Ru	45 Rh	46 Pd	47 Ag	48 Cd	49 In	50 Sn	51 Sb	52 Te	53 I	54 Xe
6s / 5d / 6p	55 Cs	56 Ba	57 La	72 Hf	73 Ta	74 W	75 Re	76 Os	77 Ir	78 Pt	79 Au	80 Hg	81 Tl	82 Pb	83 Bi	84 Po	85 At	86 Rn
7s / 6d	87 Fr	88 Ra	89 Ac															

IIIA

4f: 58 Ce, 59 Pr, 60 Nd, 61 Pm, 62 Sm, 63 Eu, 64 Gd, 65 Tb, 66 Dy, 67 Ho, 68 Er, 69 Tm, 70 Yb, 71 Lu

— Outer sub-shells as for Ce —
Outer sub-shells as for Ce and Tb

5f: 90 Th, 91 Pa, 92 U, 93 Np, 94 Pu, 95 Am, 96 Cm, 97 Bk, 98 Cf, 99 Es, 100 Fm, 101 Md, 102 No, 103 Lw

— Outer sub-shells as for Pa —

Principal shell

Sub-shell

Total number of electrons required to fill each sub-shell

1	2	3	4	5	6	7
s	s . p	s . p . d	s . p . d . f	s . p . d . f	s . p . d	s .
2	2 . 6	2 . 6 . 10	2 . 6 . 10 . 14	2 . 6 . 10 . 14	2 . 6 . 10	2

structure, in the case of the first shell). Elements of group IA, the alkali metals, possess 1 electron in their outermost shell and they will tend to lose this electron in chemical reactions exposing an inner shell containing 8 electrons (2 electrons in the case of lithium). Similarly, elements of group IIA will tend to lose 2 electrons in chemical reactions. The group VIIB elements, the halogens, all possess 7 outer-shell electrons and need to capture one additional electron in a chemical reaction.

The *valency* of an element is the number of outer-shell electrons involved in chemical reactions and is usually 1, 2, 3, or 4. An element with 6 outer-shell electrons, such as oxygen, does not have a valency of 6, but rather a valency of 2, since it can accept 2 electrons from some other element in order to obtain an outer-shell configuration of 8 electrons. The valency of an element is, therefore, equal to n or $(8 - n)$ where n is the number of outer-shell electrons, except in the case of the transition elements (see section 1.6). The elements of the transition series may show more than one valency. For example, copper may show a valency of 1 or 2 while iron may have a valency of 2 or 3.

Materials are composed of large aggregates of atoms. Within an aggregate the individual atoms are bonded together in some way. Individual atoms normally have incompletely filled outer electron-shells, but in combination with other atoms there is an attempt to obtain the filled electron shells necessary for stability. There are several types of bonding possible and these can be categorised as primary bonds and secondary bonds. Primary bonds are the bonds that exist between atoms. Secondary bonds, which are less strong than the various primary types, may exist between molecules. The primary bond types are the *ionic bond*, the *covalent bond* and the *metallic bond*. These are briefly described below, but it should be emphasised that the examples quoted are idealised. In many compound materials the nature of the bonding may be complex with several types of primary bond occurring within one compound, together with secondary bonds between molecules.

1.8 The Ionic Bond

Elements in the lower groups of the periodic table would expose a completely filled electron shell if they could lose the electrons present in their incomplete outer shell. Group IA elements, the alkali metals, would only need to shed one electron, while group IIA elements would need to lose two electrons from each atom. Conversely, atoms of the elements in groups VIB and VIIB can complete their unfilled outer electron-shell by capturing one or two electrons. An element of group I, therefore, could react chemically with an element of group VIIB, resulting in an electron transfer from one atom to the other. For example, sodium ($Z = 11$) would transfer its one electron from the third principal shell, so exposing a completely filled second shell, to chlorine ($Z = 17$), giving the chlorine atom the complement of 8 electrons in its outer layer (see figure 1.2).

By losing an electron, the sodium atom has become out of balance electrically, and is said to be in an ionised state, $Na \rightarrow Na^+ + e^-$. Similarly, by gaining an additional electron, the chlorine atom has become ionised, $Cl + e^- \rightarrow Cl^-$.

An ion is a charged particle and the net charge may be positive or negative, depending whether electrons are lost or gained. The magnitude of the net charge may be one, two, three, or in a few cases, even four units, a unit being the charge

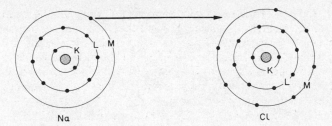

FIGURE 1.2 *Electron transfer from sodium to chlorine*

of a single electron. An ion is symbolised by using the symbol for the chemical element with a superscript indicating both the number and sign of the electrical charge.

The sodium and chlorine ions, being of opposite charge, will be strongly attracted to one another. Other forces also act upon the ions. When the ions are close to one another, there will be a repulsion between the negatively charged electron layers of both elements. There will be a gravitational attraction between the two masses. There will be one particular distance between ions at which the forces of attraction and repulsion will be equal, and the net force zero (see figure 1.3a).

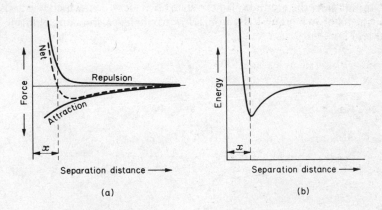

FIGURE 1.3 *Forces between adjacent atoms*

The ionic bond is strong. As will be seen from figure 1.3b, the energy level will be a minimum at the separation distance x when the net force on the ions is zero. It will require the input of considerable energy to move the ions either close to one another, or further apart.

In the solid state a large assembly of positive and negative ions will tend to form into a symmetrical arrangement, or crystal. Within a crystal each positive ion may be in contact with a number of negative ions, and vice versa. In sodium chloride, NaCl, (figure 1.4) although positive and negative ions are present in equal numbers, each sodium ion is in contact with six chlorine ions. When an ionic

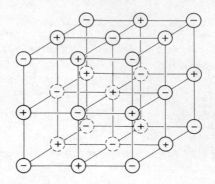

FIGURE 1.4   *Symmetrical arrangement of ions; sodium chloride* (Na^+ *and* Cl^-) *form in this pattern*

compound is melted, the crystalline arrangement ceases to exist and the ions are mobile. With mobile electric-charge carriers a molten ionic compound may conduct an electrical current and is termed an electrolyte. Ionic compounds can also dissociate to some extent when dissolved in water and ionic mobility accounts for the electrical conductivity of such solutions. Crystals are discussed in greater detail in chapter 3.

It is also possible for elements of group IIA to combine with elements of group VIIB. In this case, the group II element sheds two electrons, which will satisfy the requirements of two group VIIB atoms. For example, calcium and chlorine (see figure 1.5)

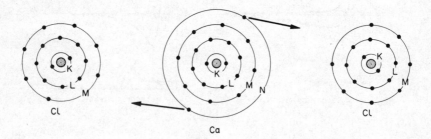

FIGURE 1.5   *Electron transfer from calcium to two chlorine atoms*

$$Ca \rightarrow Ca^{++} + 2e^-$$
$$2Cl + 2e^- \rightarrow 2Cl^-$$

giving one molecule $CaCl_2$. Similarly a group I element could combine with a group VIB element as

$$2Na \rightarrow 2Na^+ + 2e^-$$
$$O + 2e^- \rightarrow O^{--}$$

giving one molecule Na_2O.

1.9 The Covalent Bond

In covalent bonding the stable arrangement of 8 electrons in an outer shell is achieved by a process of electron sharing rather than electron transfer. In the case of chlorine for example, individual atoms combine to form diatomic molecules. The reaction can be written

$$2Cl \rightarrow Cl_2$$

The bond is achieved by the sharing of a pair of electrons. One electron from each atom enters into joint orbit around both nuclei, so giving both nuclei an effective complement of 8 outer-shell electrons (see figure 1.6a). This may be represented symbolically as

$$Cl:Cl \text{ or } Cl-Cl$$

where a pair of dots, or a dash represents a pair of electrons shared between adjacent atoms, namely one covalent bond.

In an oxygen molecule, two pairs of electrons are shared between two adjacent atoms (figure 1.6b) to give each atom a complement of eight outer shell electrons.

$$O:O \text{ or } O=O$$

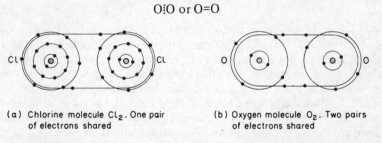

(a) Chlorine molecule Cl_2. One pair of electrons shared

(b) Oxygen molecule O_2. Two pairs of electrons shared

FIGURE 1.6 *Covalent bonds in* (a) *chlorine, and* (b) *oxygen*

The covalent bond is the form of bonding in organic molecules. (Organic chemistry is the chemistry of carbon and its compounds; organic molecules are composed principally of carbon and hydrogen.) The simplest compound of this type is methane CH_4, in which the carbon atom is bonded covalently to four hydrogen atoms thus

$$
\begin{array}{ccc}
& H & \\
& \overset{..}{\underset{..}{C}} & \\
H: & C & :H \\
& H &
\end{array}
\quad \text{or} \quad
\begin{array}{c}
H \\
| \\
H-C-H \\
| \\
H
\end{array}
$$

Not all bonds in organic molecules need be of the single covalent type. Multiple bonding is also possible. Consider the three compounds ethane, C_2H_6, ethylene, C_2H_4 and acetylene, C_2H_2.

Ethane has single covalent bonds only

$$
\begin{array}{ccc}
 & \text{H} & \text{H} \\
 & | & | \\
\text{H}- & \text{C} - & \text{C} -\text{H} \\
 & | & | \\
 & \text{H} & \text{H}
\end{array}
$$

Ethylene has a double covalent bond linking the two carbon atoms

$$
\begin{array}{cc}
\text{H} & \text{H} \\
| & | \\
\text{C} = & \text{C} \\
| & | \\
\text{H} & \text{H}
\end{array}
$$

and acetylene has a triple covalent bond between the two carbon atoms

$$
\text{H}-\text{C} \equiv \text{C}-\text{H}
$$

It should not be thought that the triple bond and double bond are stronger than a single covalent link, in fact, the reverse is true since a multiple bond is in a state of strain. A multiple bond in organic molecules can be broken more easily than a single bond. Stated in another way, this means that a compound such as ethylene or acetylene will be more chemically reactive than ethane.

The covalent bond is a strong bond. Carbon itself, in the form of diamond, illustrates this. Diamond is a highly symmetrical arrangement of carbon atoms with each atom covalently bonded to four other carbon atoms, in a tetrahedral pattern (see figure 1.7).

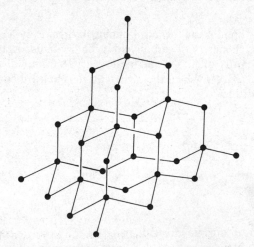

FIGURE 1.7   *Structure of diamond*

1.10 The Metallic Bond

In the bond types discussed so far, the electron distribution is rearranged to provide each nucleus with an external shell containing 8 electrons. In the case of metals, where the number of valency electrons to each atom is small, it is not possible to satisfy fully the requirement of 8 outer-shell electrons per atom, and there is only a partial satisfying of this condition. In an assembly of metal atoms, the principal forces acting will be an attraction due to gravity and a repulsion due to negatively charged electron shells in close proximity. For two atoms, the variation in energy with separation distance will be similar to that for ions (see figure 1.3) discussed earlier. (The separation distance x for minimum energy can be regarded as an atomic diameter, and for the discussion of many properties of metals, it is feasible to consider metal atoms as hard spheres of finite radius.) At the equilibrium separation distance, the outer, or valency shells of the atoms can be regarded as in contact or slightly overlapping. The outer-shell electrons, at certain points in their orbits, are attracted as much by one nucleus as by another, and the valency electrons follow complex paths (see figure 1.8). All valency

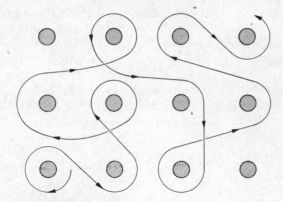

FIGURE 1.8   *Possible path of one valency electron in an array of metal ions*

electrons are shared by all the atoms in the assembly. This has similarities to a covalent bond, but the strength of the bond is weaker than the true covalent bond. One can liken the metallic state to an arrangement of positive ions permeated by an electron cloud or gas. The extreme mobility of the valency electrons accounts for the high electrical conductivity, and other properties of metals.

1.11 Secondary Bonds

In addition to the primary types of interatomic bonding discussed so far, there also exist weaker bonds, due to the polarisation of atoms or molecules. These weaker bonds, based on electrostatic attractions, are generally known as van der Waals forces.

Even the monatomic inert gases, with full outer electron-shells, will condense into liquids and solids at very low temperatures. This indicates the existence of weak bonding forces. Within such atoms, owing to the continual movement of electrons, at any instant the centroid of negative charge need not coincide with the centre of positive charge, that is, the nucleus (see figure 1.9). The atom becomes

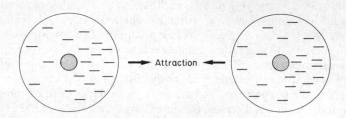

FIGURE 1.9   *Momentary uneven electron-distribution in atoms giving weak dipoles and weak interatomic attraction*

slightly polarised and may be weakly attracted to a similar polarised atom. At very low temperatures the kinetic energy of the atoms will be insufficient to overcome even these very weak attractions and the gas will condense.

The molecules of some compounds are polarised, and in such compounds there will be attraction between molecules. The attractive forces, weak in comparison with primary-bond strength, will be much stronger than those occurring in the inert gases.

In some large and complex molecules the dipole may be quite large and the magnitude of the attractive force between molecules may approach the strength of the primary bonds.

2

Chemical Principles

2.1 Stoichiometric Formulae

Elements and compounds can be expressed in symbolic form. Each of the chemical elements is designated by a simple symbol, for example, carbon, C, copper, Cu, chlorine, Cl, sodium, Na. When a compound occurs between elements the compound can be symbolised by a formula indicating the elemental constituents as for common salt, sodium chloride, NaCl.

Certain elements exist in molecular form. The gaseous elements, other than the inert gases, exist as diatomic molecules with each pair of atoms linked covalently (see section 1.9). A molecular gas such as oxygen must be symbolised as O_2, indicating that the oxygen atoms are linked in pairs.

The chemical formulae of compounds are stoichiometric, that is, they indicate the relative proportions of each element present within the compound. For sodium chloride the formula NaCl means that there is one atom of sodium for every atom of chlorine in the substance, but not that sodium and chlorine are present in equal masses (see table 2.1).

TABLE 2.1

Relationship between formula and molar mass

Substance	Formula	Relative atomic mass	Relative molecular mass	Mass of one kilomole (kg)
Copper	Cu	63.54	–	63.54
Oxygen	O_2	16.00	32.00	32.00
Sodium chloride	NaCl	Na = 22.99	58.45	58.45
		Cl = 35.46		
Water	H_2O	H = 1.008	18.016	18.016
		O = 16.00		
Methane	CH_4	H = 1.008	16.042	16.042
		C = 12.01		

The chemical formulae of simple compounds obey rules of valency, as in the following examples

   (a) compounds between elements of the same valency
      (i) monovalent elements – NaCl, LiF, KBr
      (ii) divalent elements – CaO, MgO, CaS

  (b) monovalent element combining with divalent element $-$ H_2O, Na_2S, $MgCl_2$
  (c) monovalent element combining with trivalent element $-$ $AlCl_3$, NH_3
  (d) monovalent element combining with tetravalent element $-$ CH_4, CCl_4
  (e) divalent element combining with trivalent element $-$ Al_2O_3

A variable-valency element such as copper, can form more than one compound with a second element as in the case of the two copper oxides Cu_2O (copper(I) oxide) and CuO (copper(II) oxide).

There are many simple compounds with formulae such as $MgSO_4$, KNO_3, NH_4Cl, in which the rules of valency would appear to have broken down. These are examples of compounds formed between an element and a group, or radicle. The sulphate group, SO_4^{--}, is divalent, while the nitrate group, NO_3^-, and the ammonium group, NH_4^+, are monovalent. The bonding within a group such as the sulphate group is of the type involving electron sharing, (see section 1.9) but the bond between the radicle and some other radicle or element is ionic.

2.2 Reaction Equations

When a chemical reaction occurs it can be expressed as an equation. A chemical-reaction equation is a quantitative equation. For example, when hydrogen burns in oxygen to produce water vapour, H_2O, the reaction equation is not written as

$$H_2 + O_2 = H_2O$$

but as

$$2H_2 + O_2 = 2H_2O$$

The first equation is incorrect since it does not balance properly. There are two atoms of oxygen on the left-hand side but only one on the right-hand side. The second equation is the correct one and is properly balanced. We can now rephrase this equation in several ways as follows·

  2 molecules of hydrogen + 1 molecule of oxygen = 2 molecules of water

or

  2 moles of hydrogen + 1 mole of oxygen = 2 moles of water

or

  2 kilomoles of hydrogen + 1 kilomole of oxygen = 2 kilomoles of water

Avogadro's law states that *equal volumes of all gases, under the same conditions of temperature and pressure contain equal numbers of molecules.* From this it follows that one kilomole of any gas occupies the same volume at the same conditions of temperature and pressure (one kilomole of any gas occupies a volume of 23.64 m^3 at a pressure of $101.5 \times 10^3 \text{ N/m}^2$ (1.015 bar) and at a temperature of $15°C$).

From this it follows that the equation could be written as

  2 volumes of hydrogen + 1 volume of oxygen = 2 volumes of water vapour

The equation can also be expressed in mass term

4.032 kg H_2 (2 kmol) + 32.00 kg O_2 (1 kmol) = 36.032 kg H_2O (2 kmol)

A chemical equation can also include information on the quantity of energy absorbed or emitted during the reaction, for example

$$2H_2 + O_2 = 2H_2O \qquad \Delta H = -286 \text{ MJ/kmol}$$

In this case the energy change in the reaction, ΔH, is -286 MJ per kilomole of water formed. ($\Delta H = -572$ MJ for the combustion of 2 kmol of hydrogen.) The negative sign indicates that energy is lost from the reactants and emitted to the surroundings, that is, the combustion of hydrogen is an *exothermic* reaction. If ΔH is positive it means that the reaction absorbs energy from its surroundings.

2.3 Reactivity

There are many elements that possess strong affinities for each other, but this does not necessarily mean that a reaction will take place spontaneously if the reactants are brought together. Hydrogen and oxygen have a strong affinity and can combine according to the reaction equation $2H_2 + O_2 = 2H_2O$ but at ordinary temperatures these two gases can be in intimate contact with no reaction occurring. In this and many other cases, it is necessary to increase the energy of the system in order for the reaction to proceed. The quantity of energy which has to be put into the system before the reaction will take place is termed the *activation energy* of the process. A mechanical analogy is the case of a tetragonal prism of mass m (see figure 2.1). The prism will remain in position A indefinitely if left undisturbed,

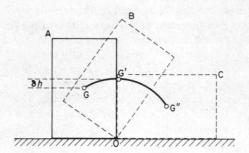

FIGURE 2.1 *Activation of a process – mechanical analogy*

even though position C is the position of lowest potential energy for the prism. If the prism is moved from position A to position C by pivoting about point O, it will be seen that the centroid of the figure will follow the path GG'G". When in position B the potential energy of the prism will be greater than the potential energy of state A by an amount $mg\delta h$. This quantity of energy would be termed the activation energy necessary for the change in prism position to occur.

The rate at which a process occurs is governed by the Arrhenius rate law, which can be written as

$$\text{rate} = A \exp \left(- \frac{q}{kT} \right)$$

where q is the activation energy, k is Boltzmann's constant, T is the temperature (K), and A is a constant. This may be rewritten as

$$\text{rate} = A \exp \left(- \frac{Q}{R_0 T} \right)$$

where Q is the activation energy per kilomole, and R_0 is the universal gas constant (8.314 kJ/kmol K).

For many chemical reactions the activation energy is of the order of 40 000 kJ/kmol. Activation energies for many physical reactions in alloy systems are much higher than this and are in the range of 150 000 to 200 000 kJ/kmol.

It will be apparent from the Arrhenius equation that a change in temperature will have an enormous effect on the rate of reaction. Consider a reaction of activation energy 150 000 kJ/kmol at two temperatures 300 K and 900 K.

At 900 K $\qquad \exp \left(- \frac{Q}{RT} \right) = \exp \left(- \frac{150 \times 10^3}{8.314 \times 900} \right)$

$$= e^{-20} \approx 10^{-8.5}$$

At 300 K $\qquad \exp \left(- \frac{Q}{RT} \right) = \exp \left(- \frac{150 \times 10^3}{8.314 \times 300} \right)$

$$= e^{-60} \approx 10^{-25.5}$$

In other words, the reaction would occur approximately 10^{17} times faster at 900 K than at 300 K.

In industrial processes catalysts are frequently used to accelerate the rate of a reaction. The presence of a catalyst effectively reduces the activation energy for a reaction. Conversely a reaction may be retarded with the aid of a negative catalyst. Catalysts and negative catalysts are not permanently consumed in the reaction.

The rate at which a reaction occurs is not solely a function of temperature, it is also affected by changes in pressure and the relative concentrations of reactants.

Consider a gaseous mixture of hydrogen and chlorine. The molecules of hydrogen and chlorine will be in a constant state of movement and the molecules will be constantly in collision. Some of these molecular collisions will be between a molecule of hydrogen and a molecule of chlorine, but not every one of these will result in a reaction to give hydrogen chloride according to the equation

$$H_2 + Cl_2 = 2HCl$$

Such a reaction will only take place when the colliding molecules possess energies in excess of the activation energy for the process, as stated in section 2.3. The rate of reaction, however, will be related to the rate at which collisions occur. The rate at which collisions occur does not depend only on temperature, but also on the relative concentrations, or partial pressures, of the reacting gases. (Similarly, for reactions that occur in liquid solutions, the rate of reaction is dependent on the concentrations of the reactants.) Therefore the reaction rate falls as the reactants are consumed in the process. A little more than a century ago the law of mass action was postulated. This law states that the rate at which a substance reacts is proportional to its 'active mass' and that the velocity of a chemical reaction is proportional to the product of the 'active masses' of the reactants. The active mass is assumed to be proportional to the molar concentrations for gases and dissolved substances, but for solid reactants the active mass is related to the surface area of the substance, and the surface condition, rather than to the total mass of substance.

2.4 Energy Change in a Reaction

When a chemical reaction takes place it is normally accompanied by a small change in energy. If the energy content of the reactants is greater than the energy content of the reaction products then excess energy is emitted as heat during the reaction. Such reactions are termed *exothermic* reactions. The heat content of a system is denoted by the symbol H, and the change in the heat content of a system in a reaction is denoted by ΔH. ΔH is regarded as negative when the total heat content of a system is decreased in the reaction. In other words, the emission of heat in an exothermic reaction is regarded as an energy loss from the system and the reaction would be written as

$$A + B = C + D \qquad \Delta H = -n \text{ J}$$

Conversely, heat is absorbed by the system in some reactions. Reactions of this type are termed *endothermic* and ΔH, the energy change in the reaction, is positive. An endothermic reaction equation would be written as

$$A + B = C + D \qquad \Delta H = n \text{ J}$$

The difference between exothermic and endothermic reactions is illustrated in figure 2.2.

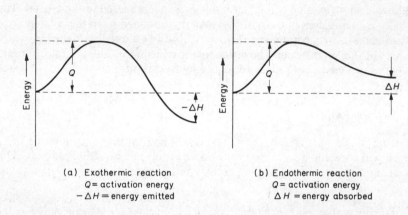

(a) Exothermic reaction
Q = activation energy
−ΔH = energy emitted

(b) Endothermic reaction
Q = activation energy
ΔH = energy absorbed

FIGURE 2.2 *Energy change in a reaction*

2.5 Combustion and Calorific Value

The reaction of a substance with oxygen is termed combustion and all combustion reactions are exothermic. Most fuels are hydrocarbon compounds and considerable heat energy is emitted when they burn in oxygen. A *calorific value*† is quoted for a fuel. For a solid or liquid fuel the calorific value is the amount of energy liberated per unit mass of fuel, with subsequent cooling of the reaction products to 15°C. For gaseous fuels the calorific value is quoted per unit volume of fuel, the volume being measured at standard conditions of temperature and pressure (15°C and a pressure of 101.5×10^3 N/m² (1.015 bar)). As in the case of solid and liquid fuels it is required that there be subsequent cooling of the combustion products to 15°C. (At a pressure of 101.5×10^3 N/m² and a temperature of 15°C, one kilomole of any gas occupies a volume of 23.64 m³.)

Consider the combustion of pure carbon. The reaction equation is

$$C + O_2 = CO_2 \qquad \Delta H = - 394.5 \text{ MJ/kmol}$$

394.5 MJ of energy are emitted during the complete combustion of 1 kmol of carbon (12.01 kg). The calorific value of the carbon is the quantity of heat energy liberated by the complete combustion of one kilogram of the element and is therefore 394.5/12.01 = 32.86 MJ/kg. For methane gas, CH_4, the combustion reaction is

$$CH_4 + 2O_2 = CO_2 + 2H_2O \qquad \Delta H = - 890 \text{ MJ/kmol}$$

890 MJ of energy are liberated during the complete combustion of 1 kmol of methane (23.64 m³). The calorific value of the methane is therefore 890/23.64 = 37.6 MJ/m³.

Methane and other fuels containing hydrogen burn to give steam as one of the combustion products. The total calorific value, as determined after cooling the reaction products to 15°C, includes the heat energy liberated as latent heat during the condensation of steam to water. This total, or 'gross', calorific value is unrealistic from a practical standpoint, since condensation of the steam and subsequent cooling does not normally occur in an industrial combustion process. A second, and lower, calorific value — the 'net' value — is often quoted in these cases. The net calorific value does not include heat energy obtained from the condensation of steam and subsequent cooling. The net calorific value for methane is about 95 per cent of the 'gross' value. The difference between gross and net calorific values is generally taken as 2440 kJ/kg for liquid hydrocarbon fuels.

2.6 Air/Fuel Ratio

Referring to the reaction $C + O_2 = CO_2$, it will be seen that one molecule of oxygen will combine with carbon to form one molecule of carbon dioxide. Therefore at standard conditions of temperature and pressure, the volume of carbon

†Heats of reaction and calorific values were formerly quoted in calories/mole and calories/gram respectively; 1 calorie = 4.187 joule. The standard unit of energy in the SI system is the joule, and the preferred unit for an amount of substance is the kilogram mole (kilomole) which is the molecular weight expressed in kilograms.

dioxide formed will equal the volume of oxygen required for the reaction. However, in the reaction $2CO + O_2 = 2CO_2$, two volumes of carbon monoxide will combine with one volume of oxygen to produce two volumes of carbon dioxide, that is, the volume of the reaction product will be only two-thirds of the volume of the reactants.

From these reaction equations it is possible to calculate the amount of air necessary for the complete combustion of a fuel.

1 kmol of carbon (12.01 kg) requires 1 kmol of oxygen (23.64 m^3) for complete combustion. Since oxygen makes up 20.9 per cent by volume of air, 1 kg of carbon requires $23.64/(12.01 \times 0.209) = 9.42$ m^3 of air (measured at 1.01 bar and 15°C) for complete combustion.

The combustion of carbon to form carbon dioxide may be shown as two separate reactions

$$C + \tfrac{1}{2}O_2 = CO \qquad \Delta H = -111.5 \text{ MJ/kmol}$$

$$CO + \tfrac{1}{2}O_2 = CO \qquad \Delta H = -283.0 \text{ MJ/kmol}$$

It will be seen from the above reactions that if there is insufficient air for complete combustion of the carbon then the combustion products will contain some carbon monoxide. Since approximately 70 per cent of the energy liberated in the total reaction is produced by the combustion of carbon monoxide it is important that sufficient air is available to ensure complete combustion. The presence of unburnt carbon monoxide in the exhaust gases constitutes a major loss of energy. In practice the combustion air supply is in excess of the theoretical amount necessary for complete combustion by some 15 or 20 per cent.

The above remarks do not apply to the combustion of fuels in internal-combustion piston engines. The fuel used in spark-ignition engines is a complex mixture of hydrocarbons, that is, compounds of hydrogen and carbon. One of the major constituents is octane, C_8H_{18}. It has been found that engine performance is best when the air/fuel ratio is 15 to 20 per cent less than that required for complete combustion, and so the exhaust gases from the engine contain a considerable proportion of carbon monoxide. The reaction for complete combustion of a hydrocarbon fuel with the general formula C_xH_y could be written as

$$C_xH_y + \left(x + \frac{y}{4}\right)O_2 = xCO_2 + \frac{y}{2}H_2O$$

The details of the combustion of hydrocarbons in internal-combustion engines are not known, but it is generally believed that it is a chain reaction involving the splitting of the hydrocarbon molecules into highly reactive intermediate compounds. Combustion in the engine cylinder should proceed rapidly but evenly. Engine 'knock' is the result of the detonation of the air/fuel mixture, that is, a virtually instantaneous propagation of combustion throughout the mixture.

2.7 Polymerisation

Living organisms have the capacity for producing large and complex molecules from simple ingredients. These large molecules are principally composed of carbon and hydrogen atoms. The long fibrous molecules produced help to determine the properties possessed by naturally occurring materials such as timber, wool, cotton and natural rubber. The whole range of plastics materials is also based upon very large molecular structures, but in this case the large molecules are produced synthetically.

Many covalently bonded compounds of carbon and hydrogen exist in nature. The simplest group of hydrocarbon compounds is the paraffin series, many members of which are found in petroleum. The first few compounds in the paraffin series are

$$
\text{methane} \quad CH_4 \quad \text{or} \quad
\begin{array}{c}
H \\
| \\
H-C-H \\
| \\
H
\end{array}
$$

$$
\text{ethane} \quad C_2H_6 \quad \text{or} \quad
\begin{array}{cc}
H & H \\
| & | \\
H-C-C-H \\
| & | \\
H & H
\end{array}
$$

$$
\text{propane} \quad C_3H_8 \quad \text{or} \quad
\begin{array}{ccc}
H & H & H \\
| & | & | \\
H-C-C-C-H \\
| & | & | \\
H & H & H
\end{array}
$$

The paraffins are known as saturated hydrocarbons since every linkage is a single covalent bond, and the carbon chain is fully saturated with hydrogen. Other hydrocarbons may be unsaturated and contain double or triple covalent bonds. An example is ethylene

$$
C_2H_2 \text{ or } \quad
\begin{array}{c}
H \\ \diagdown \\ \diagup \\ H
\end{array}
C = C
\begin{array}{c}
H \\ \diagup \\ \diagdown \\ H
\end{array}
$$

An unsaturated compound may be made to react with itself, or with other compounds, to produce large complex molecules. This is termed polymerisation. In the case of ethylene the double bonds can split allowing the separate molecules to link up forming polyethylene (polythene)

$$
\begin{array}{c}
H \ H \\
| \ | \\
C=C \\
| \ | \\
H \ H
\end{array}
+
\begin{array}{c}
H \ H \\
| \ | \\
C=C \\
| \ | \\
H \ H
\end{array}
+
\begin{array}{c}
H \ H \\
| \ | \\
C=C \\
| \ | \\
H \ H
\end{array}
=
\begin{array}{c}
H \ H \ H \ H \ H \ H \\
| \ | \ | \ | \ | \ | \\
-C-C-C-C-C-C- \\
| \ | \ | \ | \ | \ | \\
H \ H \ H \ H \ H \ H
\end{array}
$$

or

$$n \left(\begin{array}{c} H \quad H \\ | \quad | \\ C = C \\ | \quad | \\ H \quad H \end{array} \right) = \left(\begin{array}{c} H \quad H \\ | \quad | \\ - C - C - \\ | \quad | \\ H \quad H \end{array} \right)_n$$

where n is a very large number.

This is the process termed *addition* polymerisation. According to the reaction equation shown above there is a free valency electron at each end of the molecular chain. This is not permissable and there must be a small quantity of some other reactant present to provide a logical ending to the chain, for example chlorine, giving

$$\cdots \; - \underset{\underset{H}{|}}{\overset{\overset{H}{|}}{C}} - \underset{\underset{H}{|}}{\overset{\overset{H}{|}}{C}} - \underset{\underset{H}{|}}{\overset{\overset{H}{|}}{C}} - \underset{\underset{H}{|}}{\overset{\overset{H}{|}}{C}} - Cl$$

In addition-type polymers the number of atoms in each molecular chain will be extremely large, running into thousands.

Copolymerisation is the term used for the addition polymerisation of two or more simple compounds. If two monomers are represented by A and B, then copolymerisation could produce a chain molecule represented as

$$-A-B-A-B-B-B-A-$$

Another type of polymerisation reaction is termed *condensation* polymerisation. A condensation reaction is one in which two molecules combine to form a larger, and more complex molecule, together with a simple molecule, frequently water, as a by-product. The reaction between phenol, C_6H_5OH, and formaldehyde, HCHO, is an example of condensation

The nature and properties of the plastics materials produced by polymerisation reactions are discussed in later chapters.

2.8 Ionisation and Electrolysis

Ionisation is the dissociation of a compound into charged particles, or ions. Water ionises to a small extent according to the reaction

$$H_2O \rightleftharpoons H^{+}† + OH^{-}$$

This, like all ionisation reactions, is a reversible reaction. Many ionic compounds are water-soluble, and dissociate into their ions on solution.

The motion of an electrically charged particle constitutes the passage of an electrical current. The ions in an ionic solution are normally in random motion and so the net current flow in any particular direction is zero. If, due to imposed conditions, the movement is directional, then there is a definite electrical current flow taking place. Electrically conducting solutions are known as electrolytes.

If two electrodes are partially immersed in an electrolyte and there is a potential difference between them, ions will move preferentially through the electrolyte with the positively charged ions being attracted to the negative electrode, or *cathode*, and the negatively charged ions being attracted towards the positive electrode, or *anode*. The positive ions are therefore termed *cations* and the negative ions termed *anions*. When the ions reach the appropriate electrodes the charges they are carrying will be neutralised.

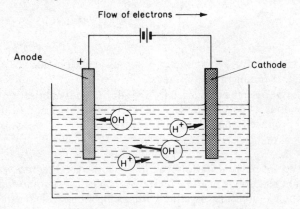

FIGURE 2.3   *Electrolytic cell showing flow of anions towards anode and flow of cations towards cathode*

In the electrolysis of water (see figure 2.3) hydrogen ions are attracted to the cathode. At the cathode there is a reaction between the excess electrons present in the cathode and the cations, liberating hydrogen

$$H^{+} + e^{-} = H \qquad (e^{-} = 1 \text{ electron})$$

At the anode the hydroxyl ions lose their charge according to the reaction

$$2(OH^{-}) = O + H_2O + 2e^{-}$$

†Strictly speaking, a hydrogen ion, H^{+}, does not exist and the dissociation of water is into hydroxyl ions, OH^{-}, and hydroxonium ions, H_3O^{+}, according to the reaction $2H_2O \rightleftharpoons H_3O^{+} + OH^{-}$, but it is customary to use the concept of the hydrogen ion, H^{+}.

The electrons produced in this reaction enter the anode and flow through the external circuit to the cathode. Thus the electrolysis decomposes the water into its constituents, hydrogen and oxygen. A point to note is that in this case one electron is associated with the formation of one atom of hydrogen, but that two electrons are associated with the liberation of one atom of oxygen, since hydrogen is monovalent but oxygen is divalent.

Faraday formulated two laws in connection with electrolysis and these may be stated as follows

(1) The amount of decomposition caused by electrolysis is in proportion to the quantity of current passed, and
(2) For the same quantity of current the amount of decomposition is proportional to the chemical equivalents of the elements liberated, the chemical equivalent being defined as, M/V where M is the atomic mass number and V the valency of the element.

The quantity of electricity required for the liberation of one atom of hydrogen is the charge of one electron, namely 1.602×10^{-19} C. The quantity of electricity required for the liberation of one gram equivalent of any element is the charge of one electron multiplied by Avogadro's number, N_A ($N_A = 6.023 \times 10^{23}$/mol — see section 1.4); that is

$$1.602 \times 10^{-19} \times 6.023 \times 10^{23} = 96\,480 \text{ C}$$

The reactions that take place at anode and cathode depend on numerous factors including the nature of the electrolyte, the nature of the electrode, and the cell voltage and current conditions. When an aqueous solution of a copper salt is electrolysed with copper electrodes, copper deposition occurs at the cathode while solution of copper takes place at the anode. This principle is utilised in the electrolytic refining of copper.

Many metals may be plated by means of electrolysis. The component to be plated is made the cathode of an electrolytic cell, and the coating deposited by electrolysis of a solution rich in ions of the coating metal. Anodising is a somewhat similar process, in which aluminium is made the anode of a cell and oxygen is 'plated 'on it, increasing the thickness of the aluminium-oxide surface film.

2.9 Corrosion

A metal in contact with an electrolyte ionises to a small extent, a typical reaction being $Zn \rightleftharpoons Zn^{++} + 2e^-$. The free electrons produced by the forward reaction remain in the metal and the ions pass into the electrolyte. Therefore at equilibrium the metal will possess a certain electrical charge. Different metals ionise to different extents and will possess different values of electrical charge when in equilibrium with their ions. The potential difference between a metal electrode and a standard hydrogen electrode is termed the *standard electrode potential*† for the metal. Table 2.2 gives the standard electrode potentials for some metals.

†Standard electrode potentials are determined with electrodes immersed in a normal solution of hydrogen ions (1 gram equivalent of hydrogen ions per litre).

TABLE 2.2
Standard electrode potentials

	Metal	Ion	Electrode potential (V)	
Base metals	Sodium	Na^+	− 2.71	Anodic
	Magnesium	Mg^{++}	− 2.38	
	Aluminium	Al^{+++}	− 1.67	
	Zinc	Zn^{++}	− 0.76	
	Chromium	Cr^{++}	− 0.56	
	Iron	Fe^{++}	− 0.44	
	Cadmium	Cd^{++}	− 0.40	
	Cobalt	Co^{++}	− 0.28	
	Nickel	Ni^{++}	− 0.25	
	Tin	Sn^{++}	− 0.14	
	Lead	Pb^{++}	− 0.13	
	Hydrogen	H^+	0.000	
	Copper	Cu^{++}	+ 0.34	
	Mercury	Hg^{++}	+ 0.79	
	Silver	Ag^+	+ 0.80	
	Platinum	Pt^{++}	+ 1.20	
Noble metals	Gold	Au^+	+ 1.80	Cathodic

All metals are electronegative; the noble metals only appear positive relative to hydrogen, the standard reference electrode.

Much metallic corrosion is of the *galvanic* or electro-chemical type. If two dissimilar metals, for example copper and zinc, are connected via an electrolyte and are also electrically connected external to the electrolyte (see figure 2.4) a

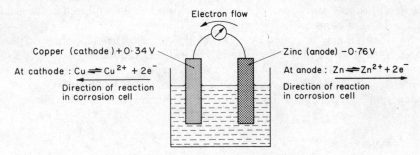

FIGURE 2.4　*Galvanic cell between copper and zinc*

galvanic cell will be set up. The potential difference of this cell will be 1.10 V, with zinc being the anode and copper being the cathode of the cell. At the anode, the reaction $Zn \rightleftharpoons Zn^{++} + 2e^-$ will be thrown out of equilibrium, as the electrons produced in the reaction flow through the external circuit to the copper, and ionisation of zinc will be accelerated in an attempt to restore equilibrium, that is, the zinc will be dissolved or corroded. Conversely, at the cathode the ionisation of copper will be suppressed due to the constant arrival of free electrons from the anode. Corrosion cells may be established in many ways in practice but in any galvanic cell the anode is corroded and the cathode protected against corrosion.

3

Aggregates of Atoms

3.1 States of Matter

Matter can be said to exist in three states, as gas, liquid and solid. The gaseous and liquid states may be classified together as the fluid state. The characteristic of a fluid, whether gas or liquid, is that the atoms or molecules which make up the material are in free movement. One kilomole of a gas at $15°C$ and atmospheric pressure occupies a volume of $23.64 \, m^3$. There are 6.023×10^{26} molecules in a kilomolecule. The diameter of many gas molecules is of the order of $2 \times 10^{-10} \, m$. From this information it can be estimated that the average separation distance between gas molecules is about $2 \times 10^{-7} \, m$, approximately one thousand times greater than one molecular diameter. A liquid, on the other hand, is a condensed state of matter and the constituent atoms or molecules are extremely close to one another. The individual molecules of a liquid are continually in motion and in collision with one another but the mean free path between successive collisions is only a small fraction of a molecular diameter, unlike the mean free path in gases at atmospheric pressure which is approximately one thousand molecule diameters.

Within a solid material the atoms or molecules are virtually in contact with each other and they occupy relatively fixed positions. The individual atoms are not completely static, and possess kinetic energy. The atomic motion, however, generally consists of small amplitude vibration about fixed positions.

3.2 Crystalline and Amorphous Solids

If a material is allowed to cool slowly from the liquid state the shape of the cooling curve obtained may be either continuous (figure 3.1a), or show discontinuities (figure 3.1b and c). As the temperature falls there is a steady decrease in the kinetic energy of the atoms or molecules that make up the liquid, and hence a steady increase in the viscosity of the fluid. Materials that show cooling curves of the type shown in figure 3.1a change during cooling from true fluids, through stages of steadily increasing viscosity, into an apparently solid condition. The atoms or molecules present still have the same type of random arrangement that existed in the true fluid state. Such materials are amorphous solids or glasses.

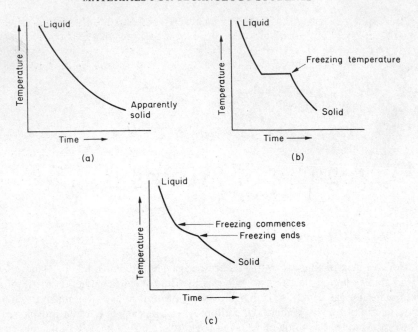

FIGURE 3.1 (a) *Cooling curve for a glassy material, no definite solidification temperature; (b) cooling curve for a pure crystalline material; (c) solidification and crystallisation occurring over a temperature range*

Other solid materials are crystalline in nature, that is, their constituent atoms or molecules are arranged in a definite symmetrical pattern. Figure 3.1b shows the type of cooling curve obtained for the freezing of a pure crystalline substance. There is a very definite freezing or solidification point, shown by a marked arrest on the cooling curve. At this temperature the atoms cease their random movement and tend to 'stick' together in relatively fixed positions in a regular pattern. Atomic motion does not cease abruptly on solidification, and the atoms or molecules in a crystalline solid vibrate about fixed positions.

The change from random fluid motion to vibration about a point within a crystal is a change to a much lower energy state. Energy, the *latent heat*, is emitted from the material at the freezing temperature. Figure 3.1c shows the type of cooling curve applicable to some crystalline mixtures. Freezing is not completed at constant temperature, but takes place over a definite temperature range. The latent-heat energy emitted during freezing shows up as a less steep gradient on the cooling curve within the freezing range.

A crystal is a symmetrical arrangement and within a crystalline solid there is a long-range orderly pattern. In non-crystalline solids and glasses there may be some symmetry or order but this will be short-range order only. As an example the structure of silica is based on the SiO_4 tetrahedron in which a silicon atom is bonded to four oxygen atoms. The oxygen atoms form the four corners of a regular tetrahedron which has a silicon atom at its centre. This is a symmetrical and ordered unit. In crystalline silica the SiO_4 tetrahedra are themselves arranged in a symmetrical pattern giving a structure with long range order but in silica glass there is no regular pattern of SiO_4 tetrahedra. The only symmetry is the short-range order within each tetrahedron (see also section 3.8 and figure 3.18).

3.3 Crystal Classes

The crystalline state is one in which the constituent atoms or molecules are arranged in a regular, repetitive, and symmetrical pattern, but a crystal will only possess a symmetrical external shape if no restraint is imposed during its growth.

A shape is said to be symmetrical if it possesses one or more of the elements of symmetry. The elements of symmetry are planes, axes, and points of symmetry. A shape of low symmetry may only possess one plane of symmetry, whereas a highly symmetrical shape, such as the cube, will contain several planes and axes of symmetry (see figure 3.2).

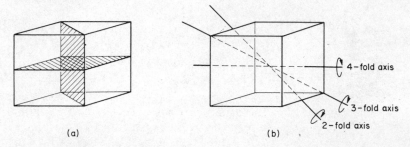

FIGURE 3.2 *Elements of cubic symmetry:* (a) *planes,* (b) *axes*

A shape is said to be symmetrical about a plane if the plane divides the shape either into two identical halves, or into two halves that are mirror images of one another. If a shape can be rotated about an axis so that the shape occupies the same relative position in space more than once in a complete reolution, then such an axis is termed an axis of symmetry. Such axes may be either 2-, 3-, 4-, or 6-fold.

There are seven crystal systems: triclinic, monoclinic, rhombohedral (or trigonal) hexagonal†, orthorhombic, tetragonal, and cubic. It is possible to define each of these systems by reference to three principal axes.

†Another system, the Miller–Bravais nomenclature, is sometimes used to define the hexagonal form. In this system reference is made to four axes, namely, three coplanar axes, a_1, a_2, and a_3, all of equal length and inclined at $120°$ to each other, and a fourth axis, of length c, perpendicular to the other three axes.

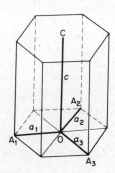

Consider three axes OA, OB and OC, of lengths a, b and c respectively, inclined to one another at angles α, β and γ (see figure 3.3). Each crystal system may now be defined in terms of the length of the principal axes, and the angles between them, as follows

triclinic	$a \neq b \neq c$	$\alpha \neq \beta \neq \gamma \neq 90°$
monoclinic	$a \neq b \neq c$	$\alpha = \beta = 90° \neq \gamma$
rhombohedral	$a = b = c$	$\alpha = \beta = \gamma \neq 90°$
hexagonal	$a = b \neq c$	$\alpha = \beta = 90°\ \gamma = 60°$
orthorhombic	$a \neq b \neq c$	$\alpha = \beta = \gamma = 90°$
tetragonal	$a = b \neq c$	$\alpha = \beta = \gamma = 90°$
cubic	$a = b = c$	$\alpha = \beta = \gamma = 90°$

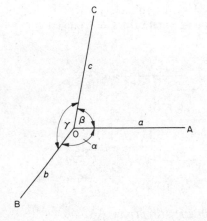

FIGURE 3.3　*Principal axes of reference*

A crystal contains myriads of atoms arranged in a regular, repetitive pattern known as a *space lattice*. The various crystal systems, as defined above, indicate the type of symmetry to be found within the space lattice, but give no indications of the relative positions of atoms within the crystal. To show this the device known as a *unit cell* is used. The unit cell can be regarded as the smallest grouping of atoms still showing the symmetry elements of the type (see figure 3.4).

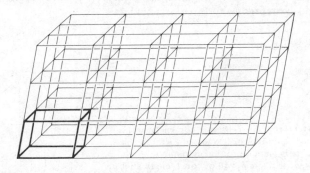

FIGURE 3.4　*Representation of part of a space lattice with a unit cell outlined*

3.4 Identification of Planes and Directions

It is sometimes necessary to refer to specific planes and directions within a crystal lattice. The notation used is the Miller indexing system. Referring to the principal axes OA, OB, and OC of the unit cell (figure 3.5), the plane PQR can be described

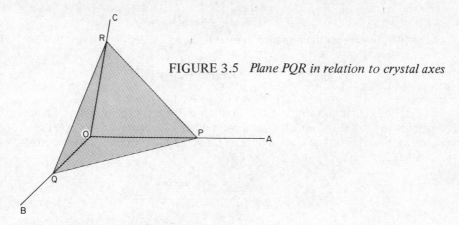

FIGURE 3.5   *Plane PQR in relation to crystal axes*

by the Miller indices *h, k,* and *l*, where these are the reciprocals of the intercepts of the plane with the principal axes, in terms of the lengths of the axes, thus

$$h \; = \; \frac{OA}{OP} \qquad k \; = \; \frac{OB}{OQ} \qquad l \; = \; \frac{OC}{OR}$$

The describing indices are enclosed in brackets, thus, (h,k,l). Any fractions are cleared to give the smallest integers.

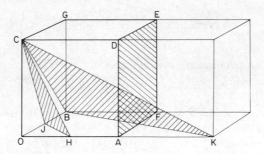

FIGURE 3.6   *Identification of planes: plane ADEF = (1,0,0); plane CJH = (2,3,1); plane BCK = (1,2,2)*

Figure 3.6 shows three planes in the cubic system. These planes would be described as follows.

Plane ADEF intercepts the principal axes at A, ∞, and ∞.

Indices of plane ADEF are $\left(\dfrac{OA}{OA}, \dfrac{OB}{\infty}, \dfrac{OC}{\infty} \right)$ or (1,0,0)

On the same basis plane CDEG would be (0,0,1), and plane BFEG would be (0,1,0). All face planes of a cube are similar and are termed a family of planes. The use of braces, thus, $\{1,0,0\}$ means reference to the whole family of face planes rather than to one specific plane.

Plane CJH intercepts the principal axes at H, J, and C.

$$\text{Indices of plane CJH are } \left(\frac{OA}{OH}, \frac{OB}{OJ}, \frac{OC}{OC}\right) \text{ or } (2,3,1)$$

Plane BCK intercepts the principal axes at K, B, and C.

$$\text{Indices of plane BCK are } \left(\frac{OA}{OK}, \frac{OB}{OB}, \frac{OC}{OC}\right) \text{ or } (\tfrac{1}{2},1,1)$$

Removal of fractions gives (1,2,2) as the indices for this plane.

To describe a direction within a crystal, draw a line from the origin of the unit cell parallel to the desired direction, and quote the co-ordinates of the point of emergence from the unit cell (see figure 3.7). Again, fractions are cleared to give the smallest integers.

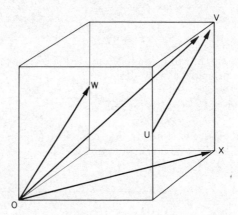

FIGURE 3.7   *Directions in crystals: direction UV* = [021] *; direction OV* = [111]*; direction OX* = [110]

Example   For direction UV draw line OW from the origin parallel to UV. Taking the length of the cube edge as unity, the co-ordinates of the points of emergence, W, are $0,1,\tfrac{1}{2}$. Removal of fractions gives the indices as 0,2,1. Indices for crystallographic directions are enclosed in square brackets, thus, [021]. Similarly, direction OV becomes [111] and direction OX becomes [110] ;

3.5 Metallic Crystals

The concept of the unit cell can be used to indicate the relative positions of atoms within a space lattice. Figure 3.8 shows three of the variations on the cubic theme, with the circles representing the positions of atom centres.

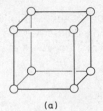 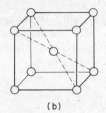 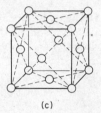

(a)   (b)   (c)

FIGURE 3.8   *Unit cells in the cubic system (circles represent atomic centres):* (a) *simple cubic;* (b) *body-centred cubic (b.c.c.);* (c) *face-centred cubic (f.c.c.)*

As stated in section 1.10 atoms and ions in the bound state can sometimes be regarded as solid spheres in contact. In the case of the crystalline form of a pure element all the spheres will be identical. However, when considering a chemical compound or a metallic alloy, atoms of two or more elements will be present, and the different elements will possess differing atomic, or ionic, diameters. In the former case, with identical atoms, the structures tend to have high symmetry. Consequently very many metals crystallise in the cubic or hexagonal forms.

Consider first uniform spheres. These may be packed together in more than one way. For example, the packing of spheres within a single plane could be square, or hexagonal (see figure 3.9) the latter being the closest possible packing for uniform spheres.

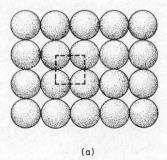 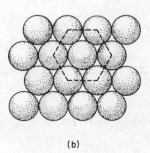

(a)   (b)

FIGURE 3.9   *Two possible packing arrangements for spheres in a plane:* (a) *square packing,* (b*) hexagonal packing*

A crystal lattice is three-dimensional, so that planes of the type shown in figure 3.9 must be stacked above one another.

A series of planes with square packing, as in figure 3.9a, if stacked upon one another so that the centres of atoms in successive layers all lie on a line normal to the base, would give a three-dimensional space lattice in the simple cubic pattern. In this system each atom is in contact with six other atoms, giving a co-ordination number of six. The spheres in a simple cubic pattern occupy 52 per cent of the available space.

In the body-centred cubic system the atoms are packed more closely than in the simple cubic form. The most densely packed planes within the body-centred cubic system are those of the (110) type (see figure 3.10). The three-dimensional

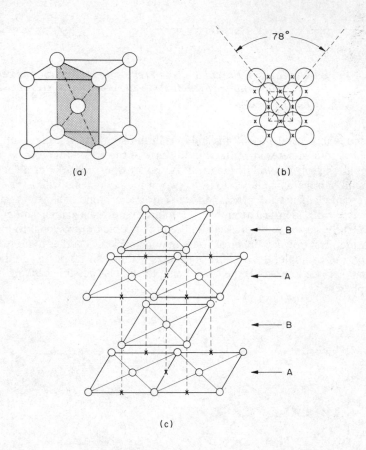

(a)

(b)

(c)

FIGURE 3.10   *Body-centred cubic system:* (a) *unit cell with* (110) *plane shaded;* (b) *plan view of a* (110) *plane, the most densely packed plane in the b.c.c. system – the positions of atomic centres in the second layer are marked x;* (c) *ABABAB-type stacking sequenc of* (110) *planes in b.c.c. structure*

crystal is built up by stacking a series of identical planes of the (110) type on one another in the manner shown in figure 3.10c. It will be seen that the atoms in alternate planes, or layers, are all in line, giving an ABABAB stacking pattern. Figure 3.13a shows the unit cell of the body-centred cubic system as a series of spheres in contact. It can be seen that the co-ordination number for this type of packing is 8. This is a higher density of packing than for simple cubic, and it can be calculated that the spheres occupy 68 per cent of the available space. A number of metals, including α-iron and chromium, crystallise in this form.

The closest possible packing of uniform spheres in a plane is hexagonal packing, and the stacking of planes of this type on one another gives rise to the closest possible packing in three dimensions. It is, however, possible to build up a symmetrical three-dimensional structure from close-packed planes in two different ways. Figure 3.11 shows the centres of atoms (marked A) arranged in a hexagonal

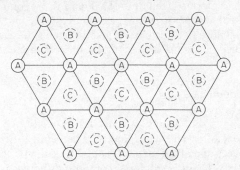

FIGURE 3.11 *Close packing of spheres; atoms in the second layer may be in either positions marked B or positions marked C*

plane pattern. When a second plane of atoms is stacked upon this first, or A, plane, the centres of atoms in the second layer may lie above either the points marked B, or the points marked C. This is also illustrated in figures 3.12a and b. If three

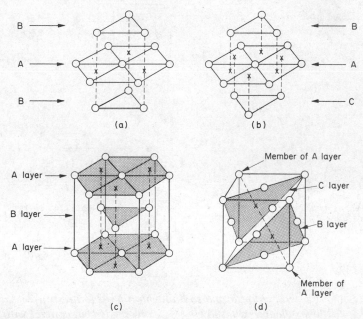

FIGURE 3.12 *Close-packed lattices:* (a) *ABAB-type stacking sequence;* (b) *ABC-type stacking sequence;* (c) *close-packed hexagonal crystal structure, ABAB stacking;* (d) *face-centred cubic crystal structure, ABCABC stacking*

relative spatial positions exist for close-packed layers then symmetry in three dimensions can be obtained if the layers are stacked in either an ABABAB sequence, or else in an ABCABC sequence. In both cases the density of packing is the same, with 74 per cent of the total space filled with spheres, and in both cases the co-ordination number is 12. The first type of stacking sequence produces the crystal form known as close-packed hexagonal, and the second sequence gives rise to the face-centred cubic type of crystal (see figures 3.12c and d).

Many metals crystallise in these close-packed systems, including magnesium and zinc (close-packed hexagonal), and aluminium and copper (face-centred cubic).

3.6 Crystalline Compounds

Compounds are composed of atoms, or ions, of more than one type. In consequence there is a great variety of crystal types that may be formed. Consider the simple case of compounds between the alkali metals (group IA of the periodic table) and the halogens (group VIIB). These are all ionic compounds between monovalent elements, for example LiF, NaCl, CsCl and KBr. These compounds all form

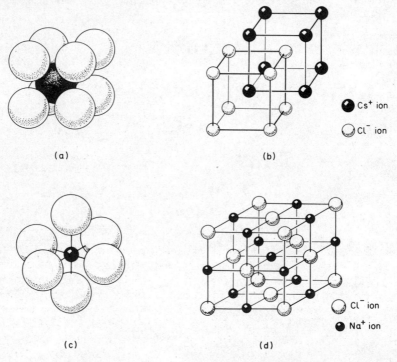

(a)

(b)

Cs$^+$ ion

Cl$^-$ ion

(c)

(d)

Cl$^-$ ion

Na$^+$ ion

FIGURE 3.13   *Caesium chloride and sodium chloride structures:* (a) *caesium chloride unit cell showing* Cs$^+$ *ion (shaded) in contact with eight* Cl$^-$ *ions;* (b) *interpenetration of cubes showing* Cl$^-$ *ion in contact with eight* Cs$^+$ *ions;* (c) *sodium chloride structure showing* Na$^+$ *ion (shaded) surrounded by six* Cl$^-$ *ions;* (d) *cubic crystal of sodium chloride*

symmetrical lattice patterns, but they do not all form in the same pattern, as one might imagine. In the case of caesium chloride, CsCl, the diameters of the caesium and chlorine ions are comparable in size. The caesium chloride structure is a body-centred cubic pattern (figure 3.13a) with each caesium ion in contact with eight chlorine ions, and vice versa. In the case of sodium chloride, NaCl, the sodium ion is small in comparison with the chlorine ion and there is only room for six chlorine ions to surround and be in contact with a sodium ion, as in figure 3.13c. This arrangement also builds up into a symmetrical cubic pattern (figure 3.13d) but of a different type from caesium choride.

Calcium fluoride, known also as fluorite or fluorspar, is an example of an ionic compound between a divalent metallic ion and a monovalent halogen; it has the formula CaF_2. The crystalline arrangement can be regarded as a face-centred cubic arrangement of calcium ions with a superimposed simple cubic pattern of fluorine ions (see figure 3.14). There are a number of other compounds that possess this type of crystalline arrangement.

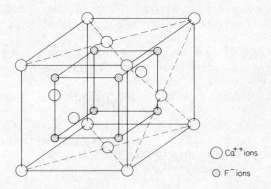

$\bigcirc$ Ca^{++} ions

$\odot$ F^- ions

FIGURE 3.14   *The structure of fluorite (calcium fluoride)*

Silicon dioxide, or quartz, SiO_2, is a covalent compound. Because of the covalent bonding it possesses great stability, as shown by its high hardness and high melting-point. The quartz crystal can be regarded as being made up of SiO_4 tetrahedra. Each silicon atom lies at the centre of a regular tetrahedron with oxygen atoms marking the four corners of the figure. Each oxygen atom is shared between two tetrahedra, and so the overall chemical composition equates to SiO_2. The size of the silicon atom is small in comparison with oxygen and the silicon fits easily into the space between oxygen atoms, giving a comparatively open structure. There are several allotropic forms of silica, since the SiO_4 tetrahedra may add together in more than one way.

The structure of silica itself forms the basis for many other compounds, the complex silicates. The open nature of the silica lattice will allow other small ions to be accommodated. Many naturally occurring minerals are complex silicates, and some of these are discussed in the next section.

3.7 Ceramics and Silicates

The term ceramic effectively covers all non-metallic, but crystalline materials. The newer industrial ceramics are principally simple oxides, carbides and nitrides. As compounds with comparatively simple formulae they tend to give crystal lattices with a high density of packing. The actual crystal form shown by each compound depends on the relative atomic diameters of the elements involved and the formula of the compound, but many of these ceramics crystallise as cubic or other high-symmetry structures. The form of interatomic bonding in these oxides, carbides and nitrides is largely covalent, giving rise to stable compounds of high hardness and high melting-point.

Some oxides exhibit the type of crystal lattice termed the *spinel* structure. The spinels are oxides with the general formula AB_2O_4, where A and B denote metals, one being divalent and the other trivalent. The unit cell of the spinel structure contains thirty-two oxygen atoms in a cubic pattern with the metal atoms located in void spaces between oxygen atoms. One of the oxides of iron forms in the spinel structure; this is magnetite, Fe_3O_4. (One of the three iron atoms is divalent and the formula could be expressed alternatively as $FeO.Fe_2O_3$.) Magnetite is magnetic (the lodestone of early navigators). A number of other spinels are also strongly magnetic. These non-metallic magnetic compounds, known as ferrites, find use in electrical and magnetic applications.

The older ceramic materials such as kaolin (china clay) and other clays are mainly silicates of complex composition. The bonding in silicates is part ionic and part covalent. Crystal-lattice structures tend to be relatively open and are often of low symmetry. Some compounds form layer lattices giving rise to thin plate-like crystals.

The structure of silicates are based on the SiO_4 tetrahedra mentioned in section 3.6. The SiO_4 unit has a deficiency of four electrons and ionic or covalent bonds are formed between the corner oxygen atoms and other tetrahedra or metal ions. The build-up of tetrahedra may give rise to a fairly open network. Figure 3.15 is a

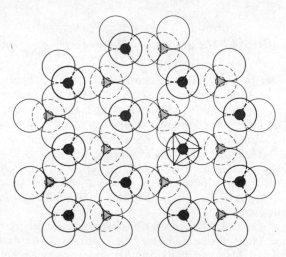

FIGURE 3.15   *Plan view of a possible silica structure*

plan view of a possible build-up of SiO_4 tetrahedra to give a three-dimensional structure. It is possible for substitution to occur with the ions of other metallic elements taking the place of some of the silicon atoms within the structure to form a wide range of silicates.

Some silicates may have chain structures with the tetrahedra adding in one plane only (see figure 3.16). An extension of the chain type of structure could give rise to a plate or layer type of lattice. This type of layer lattice is found in clays, talc and mica.

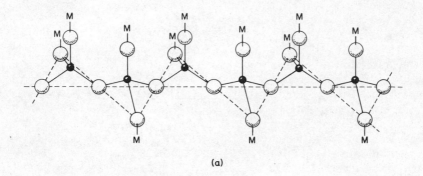

(a)

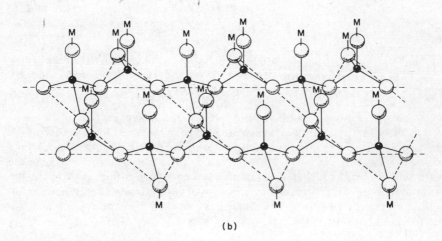

(b)

FIGURE 3.16   *Silicate chain structures:* (a) *single chain,* (b) *double chain; metal atoms may be attached to the chains at the points marked M*

In layer lattices there are strong covalent and ionic bonds within the layers, but comparatively weak van der Waals forces of attraction between the layers. This allows for easy cleavage of the crystal into thin sheets, as is the case with mica. Kaolin is a hydrated aluminium silicate and is a good example of a layer lattice (see figure 3.17). Because the structure of kaolin is not symmetrical the crystal

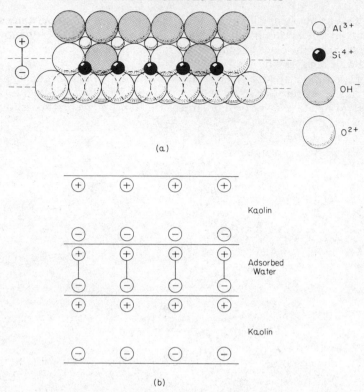

(a)

(b)

FIGURE 3.17　(a) *The structure of kaolin; due to asymmetry the kaolin layers are polarised;* (b) *adsorbed water between kaolin layers acts as lubricant giving high plasticity*

layers become highly polarised. The polarisation of the layer crystals allows the clay to adsorb water molecules, which are also polar, between the layers. This gives good intersurface lubrication and contributes to the plasticity of the clay. After a plastic clay has been shaped and 'fired' in a kiln there is a structural change in the material. First, adsorbed water is evaporated, then at high firing temperatures the water of hydration is driven off from the crystal structure. As a result new covalent bonds are created across the former interfaces between the plate-type crystals giving rise to a hard but brittle structure.

3.8 The Glass State

When heated to a high temperature crystalline silica will fuse and lose its full crystallinity. Many other crystalline materials fuse in the same manner and become amorphous. The materials in this state are not structureless, but will possess some short-range order, for example, in silica each silicon atom will still be at the centre of a tetrahedral arrangement of oxygen atoms. Rapid cooling of this amorphous material will prevent, or retard, a return to full crystallinity and preserve the amorphous structure at ordinary temperatures. This amorphous form is termed the vitreous or glass state. A two-dimensional representation of crystalline and glass structures is shown in figure 3.18.

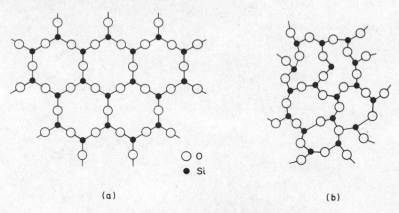

O o
● Si

(a)   (b)

FIGURE 3.18 *Representation of* (a) *crystalline, and* (b) *vitreous (glass) silica. Each silicon atom is bonded to four oxygen atoms, but for the purpose of this two-dimensional diagram silicon is shown bonded to three oxygen atoms only*

The glass state is metastable and there is a tendency for reversion to the crystalline condition. However, for many materials the activation energy of the process is such that this reversion, or devitrification, will only proceed extremely slowly, if at all, at ordinary temperatures.

Any amorphous solid material is technically termed 'glass'. The majority of the materials that are commonly termed glass and that are used for applications such as lenses, window glazing, and ordinary glassware, are the vitreous forms of complex silicates.

Many polymers may exist in the 'glass' state. At some particular temperature – the glass transition temperature, T_g – the properties of many polymer materials suffer a drastic change. At temperatures above T_g the material is in the 'rubber' state and is plastically deformable. At temperatures below T_g the material is in the glass state and is hard and brittle. The reason for this sharp property change is that when cooled to temperatures below T_g all molecular movement ceases within the material. The value of T_g is not the same for all polymers, and it is largely dependent upon the molecular structure of the material (see section 8.11).

3.9 Defects in Crystals

The structures of crystalline materials are not perfect. In addition to the gross defects, such as cracks and porosity, which may occur within bulk materials there are also sub-microscopic defects within crystals. These defects are of the same order of size as the atoms but they may exert an enormous influence on the properties of a material. The existence of these extremely small defects was postulated long before it could be verified and the imperfections rendered 'visible' by high-resolution electron microscopy. The types of defect that occur are lattice dislocations and point defects; these defects, and their effects on the physical properties of crystalline materials are described in chapter 8.

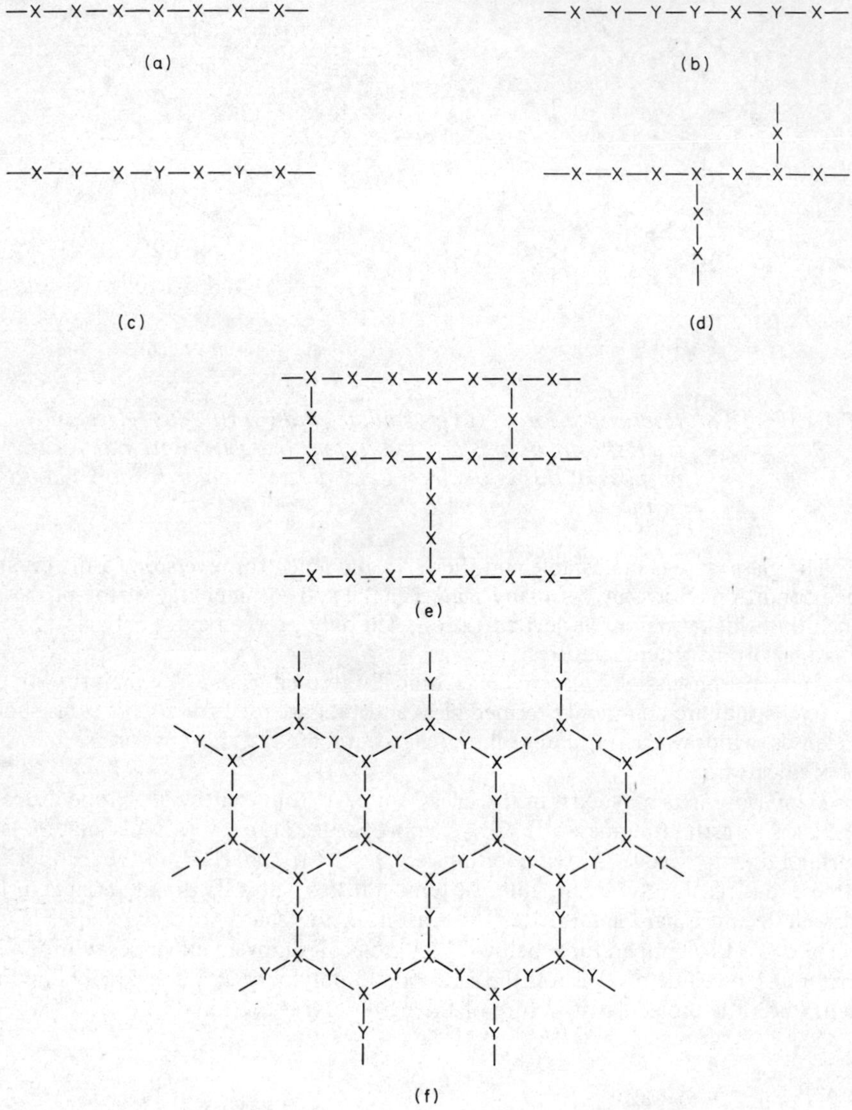

—X—X—X—X—X—X—X—

(a)

—X—Y—Y—Y—X—Y—X—

(b)

—X—Y—X—Y—X—Y—X—

(c)

$$-X-X-X-\overset{\displaystyle |}{\underset{\displaystyle |}{X}}-X-X-X-$$

(d)

(e)

(f)

FIGURE 3.19　*Polymer structure types:* (a) *linear,* (b) *random linear copolymer,* (c) *ordered linear copolymer,* (d) *branched,* (e) *cross-linked,* (f) *network (X and Y are monomers);* (d), (e) *and* (f) *are examples of non-linear polymers*

3.10 Linear and Non-linear Polymer Structures

Polymers are very large molecular structures made by reacting together small and comparatively simple molecules. The reactions are termed polymerisation processes and may be of the simple addition type, addition copolymerisation, or condensation polymerisation. The structure of the final compound and hence, to a large extent, its properties will be determined by the structure of the original reactants and the type of polymerisation process.

On the basis of structure we may sub-divide polymers into two major groupings – linear and non-linear. Each of these two groups may be further sub-divided as shown in figure 3.19. Many addition polymers are of the linear type. The term linear is really a misnomer, as the molecular chains are far from straight. The bond angle between adjacent carbon atoms is $109.5°$ and so the chain molecules become fibres of irregular shape. The use of the word linear to describe a polymer molecule signifies that there are no side chains or branches in the molecule. If a branch chain is formed the polymer becomes non-linear. Branch chains may be formed in some polymers in the presence of a catalyst. The presence of branch chains causes an increase in the strength and rigidity of the material because there is a greater chance of entanglement between molecules. If the polymer molecules become cross-linked the hardness and rigidity are increased still further. One example of cross-linking is the vulcanisation of rubber. During the vulcanisation reaction short chains of sulphur atoms form cross-links between adjacent molecules of the rubber increasing the hardness and rigidity. The degree of vulcanisation can be carefully controlled to produce the desired degree of hardness in the rubber. Fully vulcanised natural rubber, with the maximum number of cross-links is the hard rigid material known as 'ebonite'.

The very hard and rigid thermosetting materials, such as bakelite, possess three-dimensional network structures. From the foregoing it can be seen that the general form of a polymer will exert a major influence on mechanical properties. In addition the detailed shape of the polymer molecules, based on the structures of the monomers, plays a great part in determining the final properties of the material. For example, the polymerisation of ethylene ($CH_2=CH_2$) will give a simple chain molecule with only hydrogen atoms attached to the linear carbon chain. The molecules are flexible and so polyethylene is a soft and flexible material. Propylene ($CH_2=CH—CH_3$) is a related monomer, but the polypropylene molecule has a structure with $—CH_3$ groups attached to alternate carbon atoms in the molecular chain. This gives a material that is harder, and has a greater rigidity than polyethylene. In the case of polystyrene the material is even harder and more rigid because of the bulky $—C_6H_5$ groups attached to the carbon chain (see figure 3.20).

The long fibrous molecules in a molten linear polymer are in motion, but due to the extremely large average size of molecules, each molecule is not able to move freely. The motion largely takes the form of bond rotation, that is, one portion of a molecule moving in relation to another portion due to the rotation of a covalent linkage between two carbon atoms. The molecules appear to wriggle and the polymer as a whole may be thought of as a random mass of interwoven wriggling fibres. There may be weak van der Waals attractive forces between molecules. During cooling, when the kinetic energy of the molecules, and hence the extent of 'wriggling' decreases, the van der Waals forces may be strong enough to bind

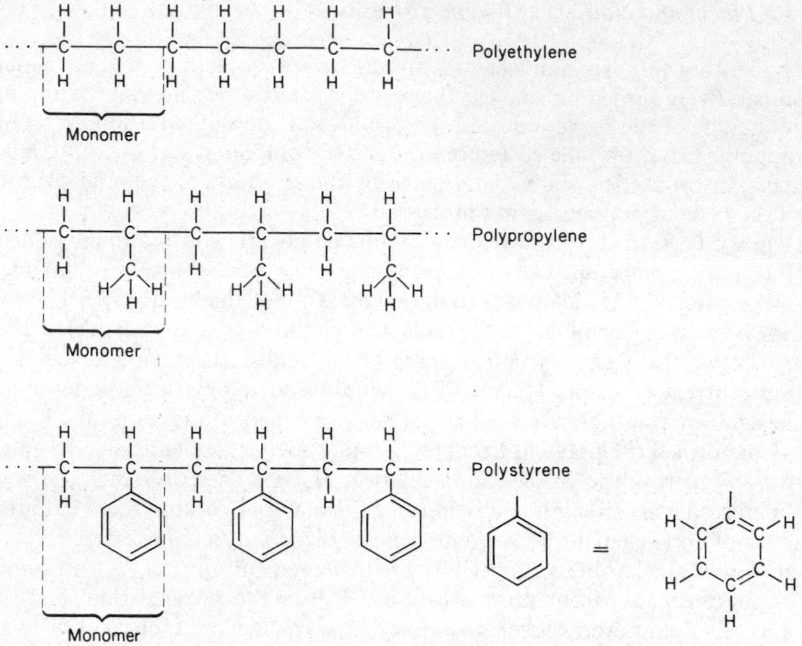

FIGURE 3.20   *Structures of three linear polymers; bulky side-groups give the material greater rigidity*

portions of adjacent molecules together, or to bind adjacent folds of the same molecule together. This will give areas of order — crystallinity — within the material. The term crystallisation is used to describe this type of formation even though the polymer is not a fully crystalline solid. The areas of regularity where portions of molecules lie in ordered bundles are termed crystallites, and these areas are of a higher density than random areas (see figure 3.21).

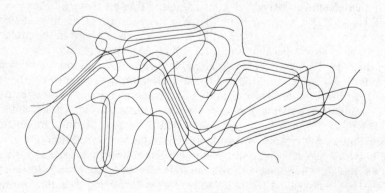

FIGURE 3.21   *Representation of crystallites and amorphous regions in a crystalline polymer*

In a polymer of simple shape, such as polyethylene, a considerable degree of crystallisation can occur. However, with molecules of more complex shape, such as the molecules of polystyrene, crystallisation does not occur, and the polymer vitrifies on cooling into the glassy state.

Some polymeric materials possess very great extensibility. Natural rubber and a number of synthetic polymers can be stretched in an elastic manner to a considerable extent. The amount of extension observed is far greater than can be accounted for by the straightening of bond angles in a 'linear' chain. Materials that show this type of elastic behaviour are termed elastomers. Within elastomers the individual linear molecules tend to coil and each molecule becomes, in effect, a minute helical spring.

Natural rubber is the compound polyisoprene. The repeating unit in the polyisoprene chain is $-CH_2-C(CH_3)=CH-CH_2-$. There are two naturally occurring materials that are composed of polyisoprene — natural rubber and gutta percha. The latter is a hard, horny material and is totally unlike rubber. The reason for the great property differences between the two materials is a comparatively minor difference in the structural formulae of the compounds. These are shown below.

```
            H                           H
            |                           |
  H  H—C—H  H    H    H  H—C—H  H    H
  |    |    |    |    |    |    |    |
 —C————C====C———C———C————C====C———C—
  |    |    |    |    |    |    |    |
  H         H    H         H
```

cis-polyisoprene (natural rubber)

```
            H                           H
            |                           |
  H  H—C—H       H    H  H—C—H       H
  |    |         |    |    |         |
 —C————C====C———C———C————C====C———C—
  |    |    |    |    |    |    |    |
  H         H    H    H         H    H
```

trans-polyisoprene (gutta percha)

In compounds such as polyisoprene containing some double covalent bonds there are two vacant spaces associated with each double bond. In the cis form both vacant spaces occur on the same side of the polymer chain while in the trans form the vacant spaces occur on opposite sides of the chain. The molecules of cis-polyisoprene may coil and so form a highly extensible material. Molecules of the trans form do not coil and the result is a hard rigid material.

The terms 'thermoplastic' and 'thermosetting' are used in connection with polymeric materials. A thermoplastic material is one that becomes more plastic when heated and becomes rigid again on cooling. Polymers with linear structures and also those that are branched or lightly cross-linked are thermoplastic. A thermosetting material is one that on first heating becomes plastic and can be moulded, but almost immediately chemical reactions occur that either complete the polymerisation of a part-polymerised material to give a full network structure, or else cause extensive cross-linking to occur. In either case the result is a hard, rigid material. Once this setting or 'curing' has taken place the material cannot again be made plastic. Some materials are cold-setting. This usually involves mixing the polymer resin with a 'hardener'. Curing takes place over a short period of time at ordinary temperatures to produce a hard and rigid network structure.

4

Elastic Behaviour

4.1 Direct Stress and Strain

When some external load is applied to a material it will be resisted by the cohesive forces between the atoms or molecules within the material. This will establish an internally acting force to balance the externally applied load. The intensity of the internal force, that is, the magnitude of the total force within a section divided by the area of the section, is termed the *stress*. On the assumption that the force is uniformly distributed across the section, the stress, σ, within the material is given by

$$\sigma = \frac{F}{A}$$

where F is the external force and A is the area of the section. The SI unit of force is the newton (N) and the unit of stress (force per unit area) is newton per metre^2 (N/m^2).

For a force of 75 newtons acting on a bar of section 15 mm by 10 mm the stress within the material of the bar is

$$\sigma = \frac{75}{10 \times 15 \times 10^{-6}}$$

$$= 5 \times 10^5 \, N/m^2 \text{ or } 0.5 \, N/mm^2$$

The types of stress normally considered are tensile, compressive, and shear stresses. When a material is in a state of stress its dimensions will be changed. A tensile stress will cause an extension of the length of the material, while a compressive stress will shorten the length. A tensile or a compressive stress acting in line is termed a direct stress. A shear stress imparts a twist to the material. The dimensional change caused by stress is termed *strain* (see figure 4.1). In direct tension or compression the strain is the ratio of the change in length to the original length. As a ratio, strain has no units and is simply a numerical value.

If a bar of length l is subject to a tensile load and the elongation produced is δl then the direct strain, ϵ, is given by

$$\epsilon = \frac{\delta l}{l}$$

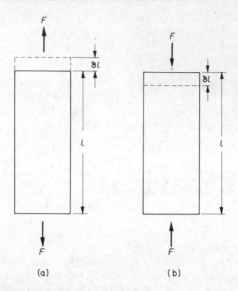

FIGURE 4.1   *Stresses and strains:* (a) *tensile force F, cross-sectional area A, tensile stress σ = F/A, tensile strain ε = δl/l;* (b) *compressive force F, cross-sectional area A, compressive stress σ = − F/A, compressive strain ε = − δl/l*

Conventionally, a direct tensile strain is deemed to be positive and a direct compressive strain is considered negative.

4.2 Elasticity

In elastic behaviour the strain developed in a material, when the material is subjected to a stress, is fully recovered immediately the stress is removed. Some materials show elastic properties up to quite high levels of stress while others possess little, if any, elasticity.

In 1678 Hooke enunciated his law, stating that the strain developed is directly proportional to the stress producing it. This law holds, at least within certain limits, for most materials. Figure 4.2 is a force-extension diagram for a metal stressed in tension. The first portion of the curve, OA, shows that the length of the specimen increases in direct proportion to the applied load, hence strain will be proportional to the applied load, and hence strain will be proportional to stress. Hooke's law does not hold beyond point A. Behaviour within the region OA will be elastic. Beyond point A the extension of the material ceases to be wholly elastic and some permanent strain is developed. The non-elastic permanent strain is termed *plastic strain.*

If the material were loaded from O to point B and the load then removed, the material would not revert to its original length, but would follow the unloading path BC showing a permanent extension, or plastic deformation, of OC when completely unstressed. Point A is known as the *limit of proportionality* or *elastic limit.*

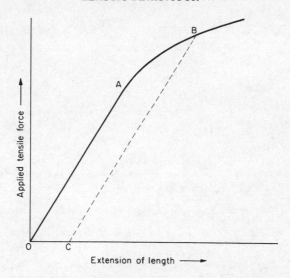

FIGURE 4.2   *Force-extension curve for a metal*

For a material that obeys Hooke's law the ratio of stress to strain, measured within the proportional range, will be a constant for the material. For direct stresses acting in tension or compression

$$\frac{\text{direct stress }(\sigma)}{\text{direct strain }(\epsilon)} = E \text{ (a constant for the material)}$$

Strain, being the ratio of change in dimension to original dimension, has no dimensions, therefore the constant E has the dimensions of stress, namely newtons/metre². E is termed the *modulus of elasticity* or *Young's modulus*, and is a measure of the stiffness of a material.

If the relationship $\sigma/\epsilon = E$ is rewritten as $\sigma/E = \epsilon$ it will be seen that in order to develop a direct strain of unity the stress within the material must equal the elastic modulus; or, in other words, the modulus of elasticity of a material can be regarded as the level of stress necessary to produce unit elastic strain in the material (that is, a doubling of the length if in tension) if such elastic deformation were possible. A material with a high value of elastic modulus will resist elastic deformation, and can be considered stiff, in comparison with a material possessing a low value of elastic modulus. To illustrate this the modulus of elasticity values for most metals lie within the range of 40 to 400 GN/m², while values for plastics materials are within the range 0.7 to 3.5 GN/m². Plastics materials are very much less rigid than metals and can be elastically strained to a considerable extent by the application of comparatively small forces.

4.3 Shear Stress and Shear Strain

Shearing forces will impart a twist to a material. If the applied forces, F, acting on a body are not in line but are equal and opposite parallel forces, the result will be to shear or twist the material, as shown in figure 4.3a. The mean value of *shearing*

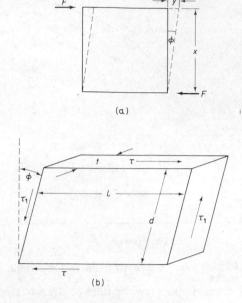

(a)

(b)

FIGURE 4.3   (a) *Shear stress and shear strain; shear force F, cross-sectional area A, shear stress* $\tau = F/A$, *shear strain* $\gamma = \tan \phi = y/x$; (b) *complementary shear stress*

stress, τ, developed within the material will be given by $\tau = F/A$ where A is the cross-sectional area of the material (measured parallel to the forces F). The units for shear stress are newton per metre^2 as in the case for direct stresses.

Shearing stresses within a material are accompanied by complementary shearing stresses. Consider a small block of material of dimensions l, d, and t, with shearing stresses τ acting on the upper and lower planes (see figure 4.3b). These shearing stresses will create a couple, of magnitude $(\tau l t)d$ and this couple must be balanced by shearing stresses τ_1 acting on the end faces of the block. The balancing couple is $(\tau_1 dt)l$

$$(\tau l t)d = (\tau_1 dt)l$$

then

$$\tau = \tau_1$$

Shearing stresses and complementary shearing stresses are of equal magnitude.

A shear stress causes a *shear strain* ϕ or γ, as in figure 4.3a. The shear strain developed is the angle of twist ϕ, in radians. But, for small strains $\phi = \tan \phi = y/x = \gamma$. Shear strain, like direct strain, is dimensionless. There is a relationship between shear stress and elastic shear strain, similar to the relationship between direct stress and direct strain

$$\frac{\text{shear stress } (\tau)}{\text{shear strain } (\gamma)} = G$$

G is a second elastic constant for a material and it is termed the *modulus of rigidity*. The units for the modulus of rigidity are also the same as those for stress, namely N/m^2. The value of the modulus of rigidity for a material is equal to that shear stress which would cause an angle of twist equal to one radian if the material could deform elastically to that extent.

4.4 The Elastic Limit

Elastic Hookean behaviour is shown by most materials, but only up to a certain limit. Referring to the force–extension curve (figure 4.2), elastic fully recoverable strain is only shown for load applications within the range OA. Consequently, point A is referred to as the elastic limit. However, the point at which the first portion of a force–extension curve ceases to be linear cannot always be determined with exactitude. When very sensitive strain-measuring devices are used the position of point A appears closer to the origin than when instruments of low sensitivity are used.

The type of force-extension curve shown in figure 4.2, is obtained during the tensile testing of many metallic materials, but a few metals, the notable one being annealed or normalised mild steel, give force-extension curves showing a sharp discontinuity or *yield point* (see figure 4.4). In these cases the material behaves

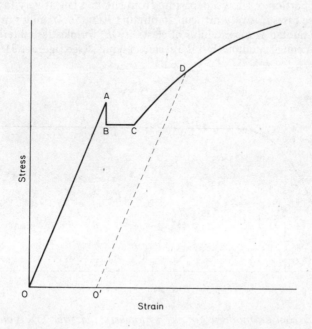

FIGURE 4.4   *Load–extension curve showing sharp yield point (mild steel)*

elastically up to the upper yield point (point A in the diagram) and it then suddenly extends in a non-elastic manner at the same, or at a lower value of applied load. If this yielding will continue at a lower value of load than the upper yield point, as in figure 4.4, point B is termed the lower yield point. After the sudden yield has occurred an increase in the applied force will cause non-proportional and non-elastic extension (portion CD in figure 4.4).

Typical force–extension curves for thermoplastic materials are illustrated in figure 8.11 and the various behaviours of polymers are discussed in section 8.11

The value of the modulus of elasticity for a material may be determined from the results of a tensile or compressive test. The slope of the first portion of the force–extension curve is related to the modulus of elasticity. The slope of the curve is $F/\delta l$ where F is the force and δl is the extension produced by force F. The stress $\sigma = F/A$ where A is the cross-sectional area of the test piece and the strain $\epsilon = \delta l/l$ where l is the original testpiece length. Young's modulus, E is given by stress/strain, therefore

$$E = \text{slope of force–extension curve} \times \frac{l}{A}$$

The modulus of rigidity for a material may be similarly determined from the results of a torsion test on a bar specimen of circular cross-section where a curve of force against angle of twist is plotted. The general shape of a force–angle-of-twist curve obtained in a torsion test is similar to the shape of a force–extension curve for the same material.

The force–extension curves for many plastics materials either show no initial straight-line portion or show a departure from Hooke's law at very low values of load. In these cases it is very difficult to obtain a value for Young's modulus. The value that is quoted as the modulus of elasticity, E, for plastics materials is not, in fact, a true Young's modulus but is a secant modulus (see figure 4.5)

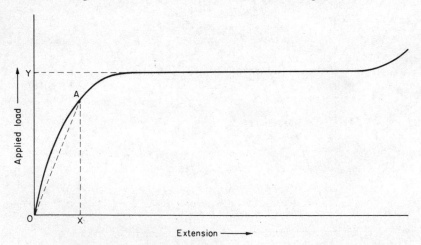

FIGURE 4.5   *Secant modulus for a thermoplastic material. OX is an extension corresponding to a strain of 0.2 per cent. The slope of the straight line OA is used to determine the modulus of elasticity (secant modulus) for the plastic*

$$E \text{(secant modulus) for plastics} = \frac{\text{stress to give strain of 0.002 (0.2\% strain)}}{0.002}$$

The amount of elastic strain that can be obtained is not the same for all materials. The maximum amount of elastic strain that can be developed in a mild steel is about 0.0013 (0.13 per cent). A high-tensile steel may be strained elastically to a greater extent and the limit is about 0.005 (0.5 per cent). Very strong 'whisker' materials, on the other hand, may take strains of up to 0.06 (6 per cent).

4.5 Relationships between Elastic Constants

The modulus of elasticity, E, and the modulus of rigidity, G, have already been mentioned. There is a third elastic modulus for a material, this being the *bulk modulus of elasticity, K*. This is the elastic response to equilateral tension, or equilateral compression (hydrostatic pressure). The value of K is given by stress/volume strain.

If a material is subjected to a direct longitudinal stress a longitudinal strain will be developed extending the length of the sample. This will be accompanied by a certain lateral contraction, compensating for the increase in length (see figure 4.6).

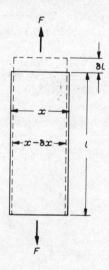

FIGURE 4.6 *Poisson's ratio: longitudinal strain = $\delta l / l$, lateral strain = $\delta x / x$,
Poisson's ratio v = lateral strain/longitudinal strain*

The ratio of lateral to axial strain is termed Poisson's ratio, v. Poisson's ratio lies between 0.28 and 0.35 for most materials.

The various elastic constants are related to one another as follows

$$G = \frac{E}{2(1 + v)} \qquad \text{or} \qquad E = 2G(1 + v)$$

$$K = \frac{E}{3(1 - 2v)} \qquad \text{or} \qquad E = 3K(1 - 2v)$$

4.6 Elastic Strain Energy

Work is performed on a material by the load applied, and when strained within the elastic range the work done is stored within the material as strain energy. The strain energy, U, is released when the deforming force is removed and the material returns to its original dimensions. For a gradually applied load the work done is given by

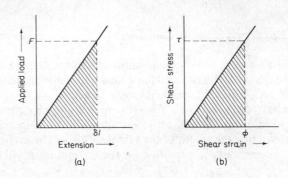

FIGURE 4.7   *Strain energy is work done (area under the curves): (a) tension, (b) shear*

the area under the force–extension curve, the shaded area in figure 4.7a. From this the strain energy, $U = \frac{1}{2}F\delta l$. Expressing this in terms of stress and the dimensions of the material, we have

$$F = \sigma A, \text{ where } \sigma \text{ is the direct stress and } A \text{ is the area, and}$$

$$\delta l = \frac{\sigma l}{E}, \text{ where } l \text{ is the total length of the testpiece}$$

Therefore

$$U = \frac{1}{2}(\sigma A)\left(\frac{\sigma l}{E}\right)$$

$$= \frac{\sigma^2}{2E} \times Al$$

But Al is the volume of the material, therefore

$$U = \frac{\sigma^2}{2E} \times \text{volume}$$

The units of strain energy, or work done, are newton metres (or joules). The strain energy per unit volume is termed *resilience* and in direct tension or compression the resilience is equal to $\sigma^2/2E$.

For shear strain the work done in straining the material is given by

$$U = \frac{1}{2}(\text{shear couple}) \times (\text{strain angle})$$

Referring to figure 4.3b the shear couple = $\tau l dt$

$$U = \frac{1}{2}(\tau l dt)\phi$$

But $\phi = \tau/G$, therefore

$$U = \frac{\tau^2}{2G} \times \text{volume}$$

This expression is comparable with that for strain energy in tension or compression.

Example Calculate the strain energy stored in a steel bolt tensioned to 6 kN. The dimensions of the bolt are shank length 25 mm, shank diameter 12 mm, threaded length 12 mm, root diameter of thread 10 mm. (E for steel $= 210\,\text{GN/m}^2 = 210\,\text{kN/mm}^2$.)

It is customary to consider that the load is distributed uniformly over the core within the threaded portion.

$$\text{Stress in shank} = \frac{6 \times 10^3}{\pi \times 6^2}$$

$$= 53\,\text{N/mm}^2$$

$$\text{Stress in threaded portion} = \frac{6 \times 10^3}{\pi \times 5^2}$$

$$= 76.4\,\text{N/mm}^2$$

Strain energy in shank ($U = (\sigma^2/2E) \times \text{volume}$) is

$$U_s = \frac{53^2 \times \pi \times 6^2 \times 25}{2 \times 210 \times 10^3}$$

$$= 19\,\text{Nmm}$$

Strain energy in thread $U_t = \dfrac{76.4^2 \times \pi \times 5^2 \times 12}{2 \times 210 \times 10^3}$

$$= 13\,\text{Nmm}$$

Total strain energy $U = U_s + U_t = 19 + 13 = 32\,\text{Nmm}$

Suppose the shank were reduced to the diameter of the thread root for the whole of its length. In this case the strain energy stored in the bolt, for the same tensile load of 6 kN would be

$$U = \frac{76.4^2 \times \pi \times 5^2 \times 37}{2 \times 210 \times 10^3}$$

$$= 40\,\text{Nmm}$$

It will be seen that the bolt with a reduced shank will store more strain energy than the other bolt for the same applied load.

4.7 Thermal Stresses

If a material is heated it will, if completely unrestrained, expand. The change in length, δl, can be calculated using the expression $\delta l = l_0 \alpha \delta \theta$ where l_0 = the original length, α = the coefficient of linear expansion, $\delta \theta$ = the temperature interval. If the material is so constrained that its dimensions cannot change and it is then heated, or cooled, through a temperature interval of $\delta \theta$ a stress will be set up within the material. The length of the sample should have been changed by an amount $l_0 \alpha \delta \theta$. The length is forced to remain as l_0 and this is equivalent to a direct strain of $(l_0 \alpha \delta \theta)/l_0 = \alpha \delta \theta$. If Young's modulus, E, for the material is known then the stress induced in the material may be calculated, the stress being $E \alpha \delta \theta$

Example   A 1000 m length of welded steel railway track is heated to 85°C before being firmly secured to the track bed. What is the stress in the track at a temperature of 15°C? E for steel = 208×10^9 N/m², α for steel = 14×10^{-6}/°C.

On cooling from 85°C to 15°C a tensile strain of $14 \times 10^{-6} \times 70$ is developed in the rail.
The tensional stress induced in the rail is

$$14 \times 10^{-6} \times 70 \times 208 \times 10^9 \text{ N/m}^2 = 203.8 \times 10^6 \text{ N/m}^2$$
$$= 203.8 \text{ MN/m}^2$$

5

Shearing Force and Bending Moment

5.1 Elastic Bending

When a bar, which is supported in some way, is subjected to a lateral force the resultant deformation of the bar is of the type known as bending. The term *beam* is used to describe a bar subjected to bending, and a beam may be considered as being *simply supported*, or *built in (encastré)*. A beam with built-in support at one end and no other support is termed a *cantilever*. Only the cases of simply supported beams and cantilevers will be considered in this book, and it will be assumed that the simple support is a knife edge. Similarly, the only types of loading that will be considered are a concentrated load (that is assumed to act at a single point) and uniformly distributed loads (U.D.L.) acting over the span of the beam, or acting over a portion of the span (see figure 5.1).

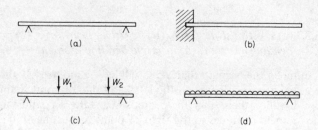

FIGURE 5.1　(a) *Simply supported beam;* (b) *cantilever;* (c) *beam with concentrated loads;* (d) *beam with uniformly distributed load (U.D.L.)*

When a beam is subjected to a force that will cause bending, both direct and shearing stresses are set up within the material of the beam and these stresses are not uniform throughout the beam. When the force tends to make the beam bend so that the upper surface is concave, the deformation is termed positive bending or *sagging*. Negative bending, or *hogging*, is said to occur when the deformation gives the upper beam surface a convex profile.

5.2 Shearing Force and Bending Moment

In a horizontal simply supported beam with concentrated loads W_1 and W_2 (see figure 5.2) there will be reaction forces R_1 and R_2 acting upwards at the supports. Consider that portion of the beam immediately to the left of the thin section X.

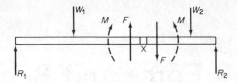

FIGURE 5.2   *Forces acting on a simply supported beam*

The forces acting on the left-hand side of the beam are load W_1 and reaction R_1, the resultant being $F = (R_1 - W_1)$. Since the beam as a whole is in equilibrium there must be an equal force F acting in the opposite direction on the right-hand side of X. The thin slice of the beam at X is therefore subjected to two equal and opposite parallel forces of magnitude F. Consequently the slice is in a state of shear and the shearing force is F. The shearing force at any point of a beam is the algebraic sum of all the vertical forces acting to one side of the point. The shearing force is considered as positive when it tends to shear the left-hand portion of a beam upwards relative to the right-hand portion. The variation of shearing force along the length of a beam may be illustrated in a shearing-force diagram.

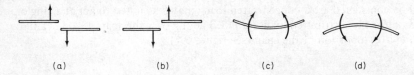

(a)   (b)   (c)   (d)

FIGURE 5.3   (a) *Positive shear-force;* (b) *negative shear-force;* (c) *positive bending-moment;* (d) *negative bending-moment*

The forces acting on the beam in figure 5.2 will have a moment M about the point X. If the moment of the forces on the left-hand side of point X about X is M in a clockwise direction then, since the beam is in a state of equilibrium, it follows that the moment of forces to the right of point X must be M in an anti-clockwise direction. The bending moment is deemed to be positive when it produces sagging of the beam, and negative when it causes hogging.

5.3 Shearing-force and Bending-moment Diagrams

Consider a simply supported beam ACB of length L and bearing a concentrated load W† applied at point C, where distance AC is a and distance CB is b (see figure 5.4). There are reaction forces R_A and R_B at the supports A and B, such

†The term load, unless otherwise stated, denotes a weight (force) acting on the beam, and as such, loads are expressed in newtons.

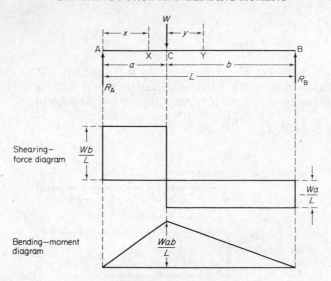

FIGURE 5.4   *Shearing-force and bending-moment diagrams for beam ACB*

that $R_A + R_B = W$. Taking moments about A

$$R_B L = Wa$$

$$R_B = \frac{Wa}{L}$$

therefore

$$R_A = W - \frac{Wa}{L} = W\left(\frac{L - a}{L}\right) = \frac{Wb}{L}$$

In the left-hand portion of the beam, AC, the shearing force F is constant and is equal to $R_A = Wb/L$. In the right-hand portion of the beam, CB, the shearing force F is constant and is equal to $R_A - W = -R_B = -Wa/L$. For the bending moment consider the forces acting to the left of a point.

The bending moment at point X will be

$$M_X = R_A x = \frac{Wbx}{L}$$

The bending moment for the section AC of the beam will increase linearly from a value of zero at A $(x = 0)$ to a maximum value of Wba/L at C, where $x = a$. Similarly, the bending moment at point Y will be

$$M_Y = R_A (a + y) - Wy$$

Over the right-hand portion of the beam, CB, this will vary linearly from a value of Wba/L at C $(y = 0)$ to a value of zero at B when $y = b$. For the special case where a concentrated load W acts at mid-span

$$a = b = \frac{L}{2}$$

therefore

$$R_A = R_B = \frac{W}{2}$$

From this it follows that the magnitude of the shearing force will be $W/2$ in the left-hand portion, and $- W/2$ in the right-hand portion of the beam. The maximum bending moment will occur at mid-span and will be given by

$$M_{max} = \frac{WL}{4}$$

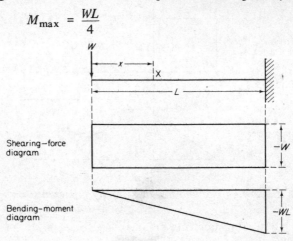

FIGURE 5.5 *Shearing-force and bending-moment diagrams for a simple cantilever*

For a simple cantilever (see figure 5.5) with a concentrated load W at the free end the shearing force F is uniform over the whole length and is equal to $- W$. The bending moment at any point X is given by

$$M_X = - Wx$$

Over the length of the cantilever this value will increase linearly from zero at the free end ($x = 0$) to a maximum value of $- WL$ at the fixed end, where $x = L$.

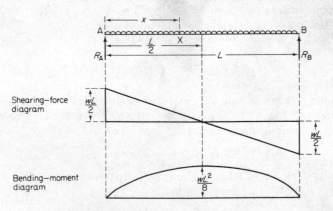

FIGURE 5.6 *Shearing-force and bending-moment diagram for a simply supported beam carrying a uniformly distributed load*

Figure 5.6 is a representation of a simply supported beam of length L carrying a uniformly distributed load of w per unit length. The total load W on the beam is given by

$$W = wL$$

and, from symmetry, the reaction at each support is $wL/2$.

Shearing force   At the support A the shearing force F is given by

$$F = R_A = \frac{wL}{2}$$

At some point X the shearing force F is given by

$$F = R_A - wx = w\left(\frac{L}{2} - x\right)$$

Thus the shearing force has a value of zero at mid-span and is equal to $-wL/2$ at the support B.

Bending moment   In order to obtain the bending moment at some point X the distributed load between A and X is considered as acting as a concentrated load at its centre of gravity. Therefore the bending moment at X, M_X, is given by

$$M_X = \frac{wLx}{2} - \frac{(wx)x}{2}$$

$$= \frac{wx}{2}(L - x)$$

The bending-moment diagram for this case is parabolic in form with a maximum value at mid-span. The maximum bending moment (when $x = L/2$) is given by

$$M_{max} = \tfrac{1}{2}\frac{wL}{2}\left(L - \frac{L}{2}\right) = \frac{wL^2}{8} = \frac{WL}{8}$$

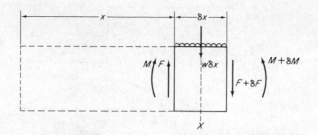

FIGURE 5.7   *Shearing-force and bending-moment relationships*

The shearing force, bending moment, and beam loading are inter-related. Consider a small segment, of length δx, of a loaded beam (this is shown in figure 5.7). If the beam carries a uniformly distributed load of w per unit length, the load on this small segment is $w\delta x$ and this load acts through the centre of the segment, X. Let the value of shearing force at each side of the thin slice be F and $F + \delta F$ respectively. Similarly, let the corresponding values of bending moment be M and $M + \delta M$. Taking moments about X for this thin segment, we have

$$M + F\frac{\delta x}{2} + (F + \delta F)\frac{\delta x}{2} = M + \delta M$$

$$F\delta x + \delta F\frac{\delta x}{2} = \delta M$$

The term $\delta F\delta x/2$ is so small it may be neglected, and so, taking the limit

$$F = \frac{dM}{dx}$$

Considering the vertical forces

$$F = w\delta x + (F + \delta F)$$

Therefore, at the limit

$$w = -\frac{dF}{dx}$$

$$= -\frac{d^2 M}{dx^2}$$

Examples with concentrated loads

1. A beam AB of length 1 m is simply supported at A and B and carries a concentrated load of 20 kN at its mid point. Determine the shearing force and bending moment diagrams for this system (see figure 5.8).

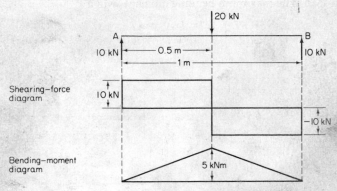

FIGURE 5.8   *Shearing-force and bending-moment diagram*

From symmetry, the reaction at each support is 10 kN. Shearing force: at all points on the left-hand side the shearing force F will be $F = 10$ kN. At all points on the right-hand side of the concentrated load the shearing force F will be

$$F = 10 \text{ kN} - 20 \text{ kN} = -10 \text{ kN}$$

Bending moment: at each support the value of the bending moment will be zero, and at the mid point of the beam the bending moment will be

$$M = 10 \times 0.5 = 5 \text{ kNm}$$

2. A beam ABCD of length 5 m is simply supported at A and D and carries concentrated loads of 30 kN and 20 kN at B and C respectively. AB = 1 m, BC = 2 m, CD = 2 m. Draw the shearing-force and bending-moment diagrams (figure 5.9).

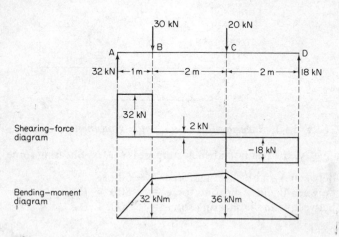

FIGURE 5.9   *Shearing-force and bending-moment diagram*

To obtain the reaction at support A, take moments about D.

$$R_A \times 5 = 30 \times 4 + 20 \times 2 = 160$$
$$R_A = 32 \text{ kN}$$
$$R_B = 50 - 32 = 18 \text{ kN}$$

Shearing force

> Between A and B the shearing force = R_A = 32 kN
> Between B and C the shearing force = $R_A - 30$ = 2 kN
> Between C and D the shearing force = $R_A - 30 - 20 = -18$ kN

Bending moment

> Bending moment at A is zero
> Bending moment at B = M_B = $R_A \times 1$ = 32 kNm
> Bending moment at C = M_C = $R_A \times 3 - 30 \times 2 = 96 - 60 = 36$ kNm
> Bending moment at D is zero

3. A beam ABCD is simply supported at B and C. There are concentrated loads, each of 500 N, at points A and D. AB = CD = 0.5 m; BC = 2.5 m. Draw the shearing-force and bending-moment diagrams (figure 5.10).

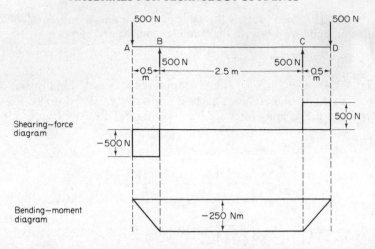

FIGURE 5.10 *Shearing-force and bending-moment diagram*

From symmetry, the reaction at each support is 500 N. Shearing force

Between A and B shearing force $= -500$ N
Between B and C shearing force $= -500 + R_B = 0$
Between C and D shearing force $= -500 + R_B + R_C = 500$ N

Bending moment

At A bending moment M_A is zero
At B bending moment $M_B = -500 \times 0.5 = -250$ Nm
At C bending moment $M_C = -500 \times 3.0 + 500 \times 2.5 = -250$ Nm
At D bending moment M_D is zero

Examples with uniformly distributed loads

1. A beam ABCD, of length 8 m, is simply supported at B and C, which are set
5 m apart. The beam carries a uniformly distributed load of 1500 N/m over its
whole length; the overhang AB is 0.6 m (see figure 5.11). To find the reactions at B
and C take moments about C.

The total force on the beam is $1500 \times 8 = 12\,000$ N and for the purpose of
determining the support reactions this force may be assumed to act at the centre
of the beam, that is, 1.6 m to the left of C.

$$R_B \times 5 = 1500 \times 8 \times 1.6 = 19\,200$$

$$R_B = 3840 \text{ N}$$

$$R_C = 12\,000 - 3840 = 8160 \text{ N}$$

Shearing force: Between A and B the shearing force varies linearly from zero at A
to a value of $-(1500 \times 0.6) = -900$ N at B. At B the shearing force changes
sharply from -900 N to a value of

$$-900 + R_B = 2940 \text{ N}$$

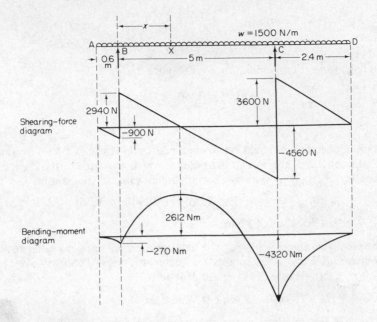

FIGURE 5.11 *Shearing-force and bending-moment diagram*

At C the shearing force is given by

$$R_B - 1500 \times 5.6 = 3840 - 8400 = -4560 \text{ N}$$

The shearing force varies linearly between B and C from a value of 2940 N at B to a value of -4560 N at C. The value of the shearing force is zero at a point situated $5 \times (2940/7500) = 1.96$ m to the right of support B. At C there is a sharp change in the shearing force from -4560 N to a value of $R_B + R_C - 1500 \times 5.6 = 12\,000 - 8400 = 3600$ N. Between C and D the shearing force decreases linearly from 3600 N at C to a value of zero at D.

Bending moment: each overhanging end of the beam acts as a cantilever. For section AB, the bending moment at the free end A is zero, while the bending moment at the support B is given by

$$M_B = - \left(1500 \times 0.6 \times \frac{0.6}{2} \right) = -270 \text{ Nm}$$

Similarly, for section CD the value of the bending moment at C is given by

$$M_C = - \left(1500 \times 2.4 \times \frac{2.4}{2} \right) = -4320 \text{ Nm}$$

and the value of bending moment reduces to zero at the free end D. Between the supports B and C the bending-moment curve is parabolic. At some point X at distance x from B the bending moment, M_X, is given by

$$M_X = R_B x - 1500(0.6 + x) \left(\frac{0.6 + x}{2} \right)$$

$$= 3840x - \frac{1500}{2}(0.36 + 1.2x + x^2)$$

$$= -750x^2 + 2940x - 270$$

The bending moment has a maximum value when $dM/dx = 0$.

$$\frac{dM}{dx} = -1500x + 2940 = 0$$

$$x = 1.96 \text{ m}$$

The maximum bending moment occurs at a point that is situated 1.96 m to the right of support B, and this is also the point at which the value of the shearing force is zero. Substituting to obtain the maximum value of the bending moment

$$M_{max} = -(750 \times 1.96^2) + (2940 \times 1.96) - 270$$

$$= 2612 \text{ Nm}$$

It will be seen from figure 5.11 that the bending moment has zero value at two points between B and C. These points are known as *points of contraflexure*. At the points of contraflexure the moment M is zero

$$-750x^2 + 2940x - 270 = 0$$

The solution of this equation gives x the values of 0.094 m and 3.83 m.

2. A cantilever ABCDE carries a U.D.L. of 800 N/m over the portion AB of its length together with concentrated loads of 10 kN and 5 kN at C and D respectively; AB = BC = CD = DE = 0.5 m. Draw the shearing-force and bending-moment diagrams for this system (figure 5.12).

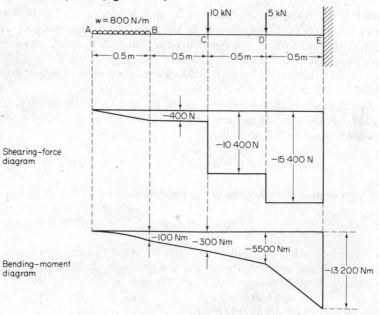

FIGURE 5.12   *Shearing-force and bending-moment diagram*

Shearing force

At A the shearing force is zero

Between A and B the shearing force varies linearly from zero to a
value of $-(800 \times 0.5) = -400$ N at B

Between B and C the shearing force is -400 N

Between C and D the shearing force is $-(800 \times 0.5) - 10\,000 = -10\,400$ N

Between D and E the shearing force is
$-(800 \times 0.5) - 10\,000 - 5000 = -15\,400$ N

Bending moment

At A the bending moment is zero

At B the bending moment is given by

$$M_B = -\left(800 \times 0.5 \times \frac{0.5}{2}\right) = -100 \text{ Nm}$$

Between A and B the bending moment varies in a parabolic manner.

At C the bending moment is given by

$$M_C = -800 \times 0.5 \times 0.75 = -300 \text{ Nm}$$

Between B and C the bending moment varies linearly.

At D the bending moment is given by

$$M_D = -(800 \times 0.5 \times 1.25) - (10\,000 \times 0.5)$$
$$= -500 - 5000 = -5500 \text{ Nm}$$

At E the bending moment is given by

$$M_E = -(800 \times 0.5 \times 1.75) - (10\,000 \times 1.0) - (5000 \times 0.5)$$
$$= -700 - 10\,000 - 2500 = -13\,200 \text{ Nm}$$

5.4 Graphical Method for Bending Moment

Figure 5.13 shows a beam simply supported at each end and carrying several
concentrated loads. This sort of problem could be solved by taking moments as
described above, but if the loading system is complex it could be quicker to solve
the problem by the following graphical technique.

Select a suitable scale and draw a load scale abcd, such that ab = W_1, bc = W_2,
and cd = W_3. From a pole O (in any position to the left of the load scale) join Oa,
Ob, Oc, and Od. This diagram is termed a *polar diagram*. Project lines downwards
from the beam supports and the points of loading. From a point p on the
projected line from reaction R_1, draw line pq parallel to Oa. Similarly, draw line
qr parallel to Ob, line rs parallel to Oc, and line st parallel to Od. Join pt and
draw line Oe in the polar diagram parallel to line pt. The lengths of the lines ae and
ed give the magnitudes of the reactions R_1 and R_2 respectively.

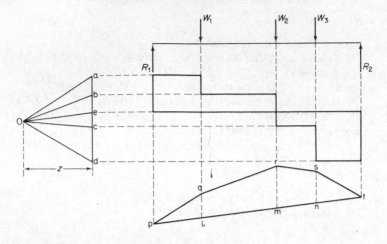

FIGURE 5.13   *Graphical method for obtaining shearing-force and bending-moment diagrams*

The diagram pqrst is a bending-moment diagram for the beam and the magnitude of the bending moment at any point on the beam is represented, to scale, by the vertical ordinate at the corresponding point of the bending-moment diagram, for example, the value of the bending moment at the point of application of the load W_2 is in proportion to the length of the line rm.

If the load scale is 1 mm $\equiv x$ N, and the scale for beam dimensions is 1 mm $\equiv y$ m, then the scale of the bending-moment diagram is given by

$$1 \text{ mm} \equiv xyz \text{ Nm}$$

where z is the distance, in mm, of the point O from the load scale in the polar diagram.

The shearing-force diagram may be obtained by projecting across from the polar diagram. The projection from point e on the polar diagram forms the base line for the shearing-force diagram.

If uniformly distributed loads are involved then, by taking small sectors and considering that within each sector the distributed load acts as a concentrated load, an approximation to the true bending-moment diagram may be obtained.

Summary

Shearing force F: the shearing force at a point in a beam is the algebraic sum of all the vertical forces acting to the left of the point.

Bending moment M: The bending moment at a point in a beam is the algebraic sum of the moments of all the forces acting to the left of the point.

Type of beam	Shearing-force diagram	F_{max}	Bending-moment diagram	M_{max}
W ↓ , L		$-W$		$-WL$
$W = wL$, L		$-W$		$-W\frac{L}{2}$
W ↓ , $\frac{L}{2}$, $\frac{L}{2}$		$\frac{W}{2}$		$\frac{WL}{4}$
$W = wL$, L		$\frac{W}{2}$		$\frac{WL}{8}$

Questions

5.1   A beam ABCD of length 5 m is simply supported at B and C such that AB is 1 m and CD is 2 m. The beam carries a uniform load along its whole length of 400 N/m. In addition there are concentrated loads of 1600 N at A and 400 N at D.
   (a) Calculate the reactions at the supports B and C.
   (b) Draw, to scale, the shearing-force and bending-moment diagrams for the beam.
   (c) From the diagrams measure the magnitude of the bending moment at the supports and at the point of change between the supports.
(*Ans.* R_B = 2500 N; R_C = 1500 N; M_B = 1800 Nm; M_C = 1600 Nm; B.M. 1.25 m from B = − 1480 Nm)

(AEB)

5.2   A beam is simply supported and loaded as shown in the figure.

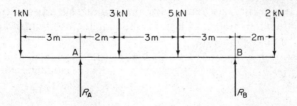

   (a) Determine the magnitude of the reactions at the supports A and B.
   (b) Draw the shear-force and bending-moment diagrams to scale showing the principal values of shear force on your diagram.
   (c) State the magnitude and position of: (i) the maximum 'hogging' moment, (ii) the maximum 'sagging' moment.
   (d) State the position of any points of contraflexure, giving their distance from the left-hand end of the beam.

(*Ans.* (a) R_A = 5 kN; R_B = 6 kN; (c) (i) 4 kNm at 11 m from L.H.S., (ii) 8 kNm at 8 m from L.H.S.; (d) 3.75 m and 10 m from L.H.S.)

(High Wycombe CT)

5.3   A loaded beam is shown in the diagram.

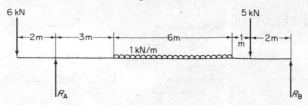

Draw the shear-force and bending-moment diagrams and sketch the deflected form of the beam. Determine the position of points of contraflexure.

(*Ans.* Point of contraflexure 4.7 m from L.H.S.)

(Bridgend TC)

5.4   A beam ABCD is 8 m long and is simply supported at B and C. AB = 1 m, BC = 5 m and CD = 2 m. The beam carries loads of 4 kN at A and 8 kN at D, together with a distributed load of 2 kN/m over the whole length. Calculate the reactions, sketch the shear-force diagram, and state the value of shear force at important points. Find the values of the bending moment at the supports, and the least value of the bending moment between the supports. Sketch the bending-moment diagram.

(*Ans.* S.F. at A = − 4 kN; at B = − 6 kN changing to 2 kN; at C = − 8 kN changing to 12 kN; S.F. = 0 at 1 m from B; B.M. = − 6 kNm at B; − 24 kNm at C; − 4 kNm)

(Middlesex Polytechnic (Hendon))

5.5   A timber roof beam ABCD has a rectangular cross-section and is 20 m long, being simply supported at two points B and C, 12 m apart. The roof imposes a uniformly distributed load of 5 kN/m over the entire length and concentrated loads of 10 kN and 20 kN are suspended from A and D respectively.

     If the points of support B and C are so positioned that the bending moments at their points of application are equal, give dimensioned sketches of the resulting shear-force and bending-moment diagrams showing the position and magnitude of the maximum bending moment.

(*Ans.* AB = 4.57 m; B.M.$_{(max)}$ = 26.1 kNm at a distance of 7.86 m from B)

(Huddersfield Polytechnic)

6

Theory of Bending

6.1 Moments of Area

Consider a thin flat plate of irregular shape (see figure 6.1a). The *centroid* of such a thin plate, or lamina, is that point at which the whole of the area of the plate may be assumed to be concentrated. The *first moment of area* of a lamina about some axis XX is given by $A\bar{y}$, where A is the total area of the lamina and $\bar{y}$ is the distance of the centroid from the chosen axis XX.

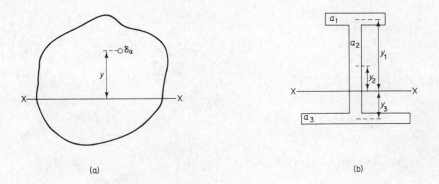

(a)   (b)

FIGURE 6.1   *First moment of area*

Consider an elemental portion of area, δa, situated at a distance y from the axis XX. The first moment of area for this element about the axis XX is $y\delta a$. The lamina can be considered as being made up of a large number of elements, $\delta a_1, \delta a_2, \delta a_3, \ldots$ situated at distances $y_1, y_2, y_3, \ldots$ respectively from the chosen axis. The first moment of area for the whole figure is then given by

$$y_1\delta a_1 + y_2\delta a_2 + y_3\delta a_3 + \ldots = \sum y\delta a$$

This summation is usually more accurately described by the integral sign $\int$
The first moment of area about axis XX is then written as

$$\int y\, da$$

The first moment of area, $\int y \, da$, will be zero when the chosen axis passes through the centroid of the lamina.

A solid body, such as a beam, can be considered as made up of a whole series of thin sections. In this book only beams of non-changing section will be discussed, that is, where the cross-sectional area is constant for any position along the length of the beam.

The shapes of beams and other components used in engineering can often be sub-divided into smaller regular figures, and the position of the centroid for a shape such as an I-beam may be readily obtained. For the I-beam cross-section shown in figure 6.1b three distinct regular areas occur — the two flanges and the web. The first moment of area for the whole section is given by

$$A_y = a_1 y_1 + a_2 y_2 + a_3 y_3$$

where a_1, a_2 and a_3 are the areas of the upper flange, web, and lower flange respectively, and y_1, y_2 and y_3 are the respective distances of the centroids of these areas from the axis XX. The distance of the centroid of the complete I-section from the chosen axis XX then becomes

$$y = \frac{a_1 y_1 + a_2 y_2 + a_3 y_3}{A}$$

Example   Determine the position of the centroid of the section shown in figure 6.2.

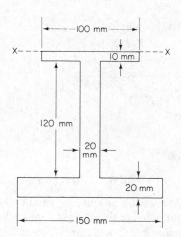

FIGURE 6.2

Area of top flange $= 10 \times 100 = 1000 \text{ mm}^2$
Area of web $= 20 \times 120 = 2400 \text{ mm}^2$
Area of bottom flange $= 20 \times 150 = 3000 \text{ mm}^2$
Total area $= 6400 \text{ mm}^2$

Let y be the distance of the centroid of the whole figure from axis XX, then

$$6400y = 1000 \times 5 + 2400 \times 70 + 3000 \times 140$$

$$y = \frac{593000}{6400} = 92.7 \text{ mm}$$

The centroid of the figure lies 92.7 mm below the top of the upper flange.

Moving bodies possess a property termed inertia, and the moment of inertia of a body about an axis is obtained by the summation of the product of particle mass and the square of its distance from the axis for all the particles making up the body. There is an analogous situation in problems concerned with beams, and an expression is required for the product of elemental areas and the square of the distance of an element from an axis for all the elements making up a total area. This is known as the *second moment of area*, I, where

$$I = \sum y^2 \delta a = y_1{}^2 \delta a_1 + y_2{}^2 \delta a_2 + y_3{}^2 \delta a_3 + \ldots$$

This summation can be described by the integral sign and the second moment of area of the section about axis XX is then given by

$$I = \sum y^2 \delta a = \int y^2 \, \mathrm{d}a$$

The second moment of area is often termed the moment of inertia, but it is incorrect to use this latter term.

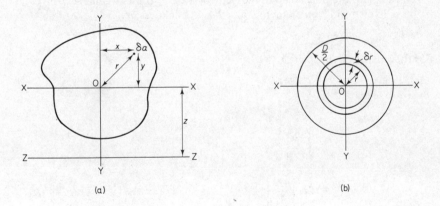

(a)   (b)

FIGURE 6.3 *Second moment of area*

The second moment of area for the figure shown in figure 6.3a about the axis XX is given by

$$I_{XX} = \int y^2 \, \mathrm{d}a$$

Similarly, the second moment of area about some other axis YY is given by

$$I_{YY} = \int x^2 \, \mathrm{d}a$$

The moment of area about some axis normal to the plane of the figure is termed the *polar second moment of area*, J. For an axis through O, and perpendicular to the plane

$$J = \int r^2 \, da$$

If the axes XX and YY are normal to each other, then $r^2 = x^2 + y^2$

$$J = \int y^2 \, da + \int x^2 \, da$$
$$= I_{XX} + I_{YY}$$

If axis XX passes through the centroid of the figure, then the second moment of area about some other axis ZZ, where ZZ is parallel to XX and situated distance z from XX, is given by

$$I_{ZZ} = I_{XX} + Az^2$$

This relationship, known as the parallel-axes theorem, may be proved as follows

$$I_{ZZ} = \int (y + z)^2 \, da$$
$$= \int y^2 \, da + 2z \int y \, da + z^2 \int da$$
$$= I_{XX} + Az^2$$

Since $\int y \, da = 0$ for an axis passing through the centroid.

To calculate the polar second moment of area, J, for a circular section (see figure 6.3b), consider an elemental area δa of thickness δr.

$$\delta a = 2\pi r \, \delta r$$

For the polar axis through O

$$J = \int_0^{D/2} r^2 \, da$$

$$= \int_0^{D/2} r^2 \, 2\pi r \, dr$$

$$= 2\pi \left[\frac{r^4}{4} \right]_0^{D/2}$$

$$= \frac{\pi D^4}{32}$$

For the circular section

$$J = I_{XX} + I_{YY}$$

and

$$I_{XX} = I_{YY}$$

$$I_{XX} = I_{YY} = \frac{\pi D^4}{64}$$

Similarly, for a hollow circular section of external diameter D and internal diameter d

$$J = \frac{\pi}{32}(D^4 - d^4)$$

and

$$I = \frac{\pi}{64}(D^4 - d^4)$$

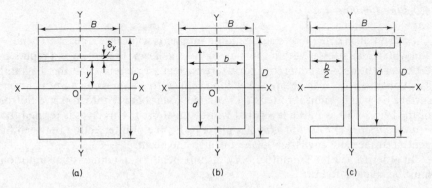

(a)   (b)   (c)

FIGURE 6.4

In the case of a simple rectangular section (figure 6.4a) for an axis **XX** passing through the centroid

$$I_{XX} = \int y^2 \, da$$

where $da = B \, dy$

$$I_{XX} = \int_{-D/2}^{D/2} y^2 B \, dy$$

$$= B \left[\frac{y^3}{3} \right]_{-D/2}^{D/2}$$

$$= \frac{BD^3}{12}$$

Similarly

$$I_{YY} = \frac{DB^3}{12}$$

For a hollow rectangular section (figure 6.4b)

$$I_{XX} = \frac{BD^3}{12} - \frac{bd^3}{12}$$

In the case of an I-beam section (figure 6.4c) the second moment of area about axis XX can be obtained in a similar manner by subtracting I_{XX} for the two rectangles of size $(b/2) \times d$ from that for the overall figure. Thus I_{XX} for the section is given by

$$\frac{BD^3}{12} - \frac{bd^3}{12}$$

For standard rolled structural sections values for I are computed graphically for the actual shape of the sections, and the results are published in reference tables.

6.2 Pure Bending

The action of shearing forces and bending moments will cause a beam to bend, and will induce stresses within the material of the beam. Figure 6.5 is a representation of a beam in bending (the curvature is greatly exaggerated in the diagram). The upper surface of the beam AA will be subject to compression and the lower surface CC will be subject to tension. At some point between the top and bottom of the beam there will be a layer that will be unstressed. This layer is termed the *neutral surface*. The *neutral axis* of a plane is the line of intersection between the neutral surface and any lateral plane through the beam.

In order to develop a simple theory of pure bending, a number of assumptions must be made as follows

(1) The beam is bent due to the action of a constant bending moment and there is no shear.
(2) The deformation of the material of the beam is wholly elastic, that is, obeying Hooke's law.
(3) The material of the beam is isotropic and possesses the same value of Young's modulus both in tension and compression.
(4) The stresses in the beam are purely longitudinal.
(5) Longitudinal surfaces and filaments bend into circular arcs with a common centre of curvature.
(6) The radius of curvature of the beam in bending is large in comparison with the dimensions of the beam.
(7) Any transverse section of the beam will remain plane, and at right angles to the neutral surface.

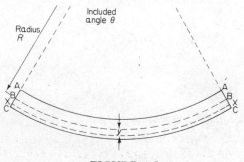

FIGURE 6.5

The neutral surface of the beam represented in figure 6.5 is denoted by line BB. For some surface XX situated at a distance y from the neutral surface, the longitudinal strain, ϵ, which is developed is given by

$$\epsilon = \frac{XX - BB}{BB}$$

$$= \frac{(R + y)\theta - R\theta}{R\theta}$$

$$= \frac{y}{R}$$

But according to Hooke's law $\epsilon = \sigma/E$ where σ = direct stress in surface XX, and E = Young's modulus. Therefore

$$\epsilon = \frac{y}{R} = \frac{\sigma}{E}$$

This may be rewritten as

$$\frac{\sigma}{y} = \frac{E}{R} \qquad (6.1)$$

E/R is constant hence *the magnitude of the stress induced at any point within a beam will be proportional to the distance of the point from the neutral surface.* Figure 6.6a is a representation of this statement. It is because the magnitude of the direct stresses developed within the material is greatest at points furthest from the neutral surface that beams of I-section are used for very many load-carrying purposes. Since the level of direct stress within the web of an I-section is considerably less than in the top and bottom flanges it becomes possible to remove some of the web material, reducing the mass per unit length of the beam, without adversely affecting the load-carrying capacity of the beam. An interesting example of this is the manufacture of the 'Castello' beam, during which a rolled section is cut and then welded in the manner shown in figures 6.6b and c.

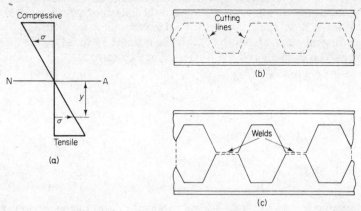

FIGURE 6.6   (a) *Distribution of direct stress within a beam section;* (b) *cutting pattern for manufacture of a 'Castello' beam;* (c) *'Castello' beam after welding*

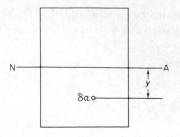

FIGURE 6.7

Consider a cross-section of a beam (as shown in figure 6.7). The neutral axis is shown as NA. For some small element δa situated at a distance y from the neutral axis, there will be a tensile stress σ acting normal to the section, that is, along the length of the beam. Since the beam is in a state of equilibrium, the tensile forces acting on one side of the neutral axis must balance the compressive forces acting on the other side of the neutral axis. The total force acting on the section of the beam is given by

$$\int \sigma \, da = 0$$

but from equation 6.1

$$\sigma = \frac{Ey}{R}$$

therefore

$$\frac{E}{R} \int y \, da = 0$$

But $\int y \, da$ is the first moment of area of the section about the neutral axis; when the first moment of area about an axis is zero the axis must pass through the centroid of the section; therefore the neutral axis contains the centroid.

The bending moment, M, which acts on the beam is balanced by the moment of the normal forces about axis NA. This may be expressed as

$$M = \int \sigma y \, da$$

But

$$\sigma = \frac{Ey}{R}$$

therefore

$$M = \frac{E}{R} \int y^2 \, da$$

But from section 6.1, $\int y^2 \, da$ is the second moment of area of the section about axis NA. I_{NA}, therefore

$$M = \frac{EI}{R}$$

where I is the second moment of area of the section about an axis containing the centroid. This equation may be rewritten as

$$\frac{M}{I} = \frac{E}{R} \qquad\qquad (6.2)$$

Combining equations 6.1 and 6.2 we have

$$\frac{M}{I} = \frac{E}{R} = \frac{\sigma}{y}$$

The above relationships are very important but it is essential, for their correct application, that the units used are consistent. Suggestions are given in table 6.1.

TABLE 6.1

	Symbol	Unit
Bending moment	M	Nm
Second moment of area	I	m^4
Modulus of elasticity	E	N/m^2
Radius of curvature	R	m
Direct stress	σ	N/m^2
Distance from the neutral axis	y	m

y is taken as positive when measured outward from the centre of curvature, and negative when measured inward for beams in positive bending (sagging). The reverse is true for beams in negative bending (hogging).

6.3 Limitations of the Pure-bending Formula

In the derivation of the bending relationships

$$\frac{M}{I} = \frac{E}{R} = \frac{\sigma}{y}$$

one of the assumptions made was that the beam was subject to a constant bending moment with no shear. In chapter 5 it was shown that the value of the bending moment varies along the length of the beam for many loading situations, and that the beam is also subject to a shearing force. However, in many cases there is zero shearing force at the point of maximum bending moment. At such a point the conditions are almost those of pure bending and the pure-bending relationships that have been derived are valid. The stresses that are developed in the material of a beam at the position of the greatest bending moment are generally the highest in the beam, and it is justifiable to use the pure-bending formula for beam-design calculations.

Example A beam of rectangular section has the following dimensions: width 75 mm, depth 50 mm, and length 1.6 m. The beam is simply supported at its ends and carries a load of 500 N at mid-span. Calculate the maximum direct stress in the beam.

From chapter 5 the maximum bending moment is $WL/4$ at mid-span

$$M_{max} = \frac{500 \times 1.6}{4} = 200 \text{ Nm}$$

From the general bending formula

$$\sigma = \frac{My}{I}$$

The second moment of area about the neutral axis for a rectangular beam is given by

$$I = \frac{BD^3}{12}$$

$$= \frac{0.075 \times 0.05^3}{12}$$

$$= 7.81 \times 10^{-7} \text{ m}^4$$

The maximum stresses will occur in the upper and lower surfaces of the beam, that is, at a distance of 25 mm (0.025 m) from the centroid. Therefore

$$\sigma_{max} = \frac{200 \times 0.025}{7.81 \times 10^{-7}}$$

$$= 6.4 \times 10^6 \text{ N/m}^2$$

Example   An I-beam of length 5 m is simply supported at each end and carries a uniformly distributed load of 20 kN/m. The dimensions of the beam are: width of flanges 150 mm, thickness of flanges 15 mm, thickness of web 10 mm, total depth of beam 200 mm. Calculate the maximum stresses developed in the material of the beam.

The second moment of area of the beam about the centroid is

$$I_{XX} = \frac{0.15 \times 0.2^3}{12} - \frac{0.14 \times 0.17^3}{12}$$

$$= 42.7 \times 10^{-6} \text{ m}^4$$

(see section 6.1.) From chapter 5 the maximum bending moment for a beam carrying a uniformly distributed load of w per unit length is $wL^2/8$ at mid-span

$$M_{max} = \frac{20 \times 10^3 \times 5 \times 5}{8}$$

$$= 62.5 \times 10^3 \text{ Nm}$$

$$\sigma_{max} = \frac{My}{I}$$

From the symmetry of the section, the centroid is at the centre of the beam, that is, 100 mm (0.1 m) from the upper or lower surfaces. The maximum direct stresses will occur at the surfaces with a maximum compressive stress being

developed in the upper surface of the beam and the maximum tensile stress
occurring in the lower surface.

$$\sigma_{max} = \frac{62.5 \times 10^3 \times 0.1}{42.7 \times 10^{-6}}$$

$$= 146.4 \times 10^6 \text{ N/m}^2$$

6.4 Deflection of Beams

When a beam is subjected to bending stresses it will be deflected. Figure 6.8 shows
the *elastic line* for a beam under load, the elastic line being the position of the
neutral surface in a deflected beam. In practice the deflection curve is almost flat
but for clarity this curve is greatly exaggerated in the figure.

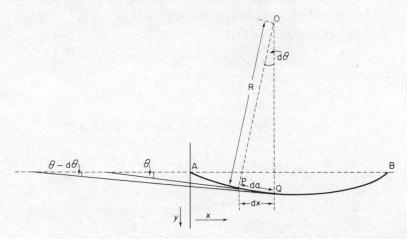

FIGURE 6.8   *Elastic line of a beam*

Beam deflection is of importance. A beam may possess sufficient strength to
sustain a load, but may deflect by too great an amount to be useful in a particular
situation. It is essential, therefore, to consider not merely the strengths of
members but also their stiffness, or resistance to deflection. The product EI is a
measure of this stiffness, and is termed the *flexural rigidity*. As shown in section
6.1 the value of I for a section is dependent not only on the dimensions of the
section, but also on the axis chosen. For a rectangular section, where $I = BD^3/12$,
such as a timber floor joist with cross-sectional dimensions of 50 mm by 200 mm,
the value of I, and hence the stiffness of the beam, is some sixteen times greater
for the condition where the beam is mounted with the larger dimension as the
depth, than it is if the smaller dimension is taken as the depth. Hence floor joists
are mounted with the larger dimension as the depth and may be used over spans
of 5 or 6 m with only small deflections under load, but are placed comparatively
close together in order that floor boards, which are mounted with their shorter
dimension as the depth, only span some 400 mm between supports.

For the case of a constant bending moment M over the whole length of a beam, the elastic line will be an arc of a circle of radius R. In most cases, however, the magnitude of the bending moment varies from point to point on the beam and, in consequence, the elastic line is not of constant radius.

Consider a small portion, PQ, of the elastic line AB in figure 6.8. It can be assumed that this small arc PQ of length da is an arc of a circle of radius R. The tangent at P makes an angle θ with the x-axis, and the difference in slope between the tangent at P and the tangent at Q is $d\theta$. The y-axis is the direction of deflection of the neutral surface. y is taken as positive downwards, that is, sagging of the beam under the action of a positive bending moment. The length of arc da is $R\,d\theta$, therefore

$$R = \frac{da}{d\theta} \quad \text{or} \quad \frac{1}{R} = \frac{d\theta}{da}$$

As one progresses along the x-axis from A toward B, θ decreases in magnitude, that is, $d\theta/da$ is negative. Taking account of signs

$$\frac{1}{R} = -\frac{d\theta}{da}$$

Since θ is very small

$$\theta = \tan\theta = \frac{dy}{dx}$$

For an almost flat elastic line, da is very nearly equal to dx, therefore

$$\frac{d\theta}{da} = \frac{d^2y}{dx^2} = -\frac{1}{R}$$

But from the general bending-equation

$$\frac{1}{R} = \frac{M}{EI}$$

$$M = -EI\frac{d^2y}{dx^2}$$

Now, if M can be expressed in terms of x, this equation can be integrated to give the slope of the beam, (dy/dx), and integrated a second time to give deflection (y). There will be integration constants, but these may be evaluated from a knowledge of the values of slope and deflection at particular points of the beam. For example, in a cantilever both slope and deflection are zero at the fixed end. For a simply supported beam with a concentrated load at mid-span, the slope is zero at mid-span, the point of maximum bending moment, and the deflection is zero at the supports.

6.5 Slope and Deflection by Calculus

Example For a beam carrying a uniformly distributed load of w per unit length and simply supported at each end the bending moment, M, expressed as a function of x is given by

$$M = \frac{wLx}{2} - \frac{wx^2}{2}$$

(from chapter 5) therefore

$$EI\frac{d^2y}{dx^2} = -M = -\frac{wLx}{2} + \frac{wx^2}{2}$$

Integrating

$$EI\frac{dy}{dx} = -\frac{wLx^2}{4} + \frac{wx^3}{6} + A$$

At mid-span, when $x = L/2$, dy/dx (slope) = 0 therefore

$$A = \frac{wL^3}{16} - \frac{wL^3}{48} = \frac{wL^3}{24}$$

Thus

$$EI\frac{dy}{dx} = -\frac{wLx^2}{4} + \frac{wx^3}{6} + \frac{wL^3}{24}$$

Integrating again

$$EIy = -\frac{wLx^3}{12} + \frac{wx^4}{24} + \frac{wL^3x}{24} + B$$

At a support, when $x = L$, y (deflection) = 0, therefore

$$B = \frac{wL^4}{12} - \frac{wL^4}{24} - \frac{wL^4}{24} = 0$$

The maximum deflection occurs at the point of maximum bending moment, that is, at mid-span, where $x = L/2$. Substituting $x = L/2$

$$EIy_{max} = -\frac{wL^4}{96} + \frac{wL^4}{384} + \frac{wL^4}{48}$$

$$= \frac{5wL^4}{384}$$

But, for a uniformly distributed load, the total load $W = wL$, therefore

$$y_{max} = \frac{5WL^3}{384EI}$$

The maximum slope occurs at each support, that is, when $x = 0$ and $x = L$. For $x = 0$

$$EI\frac{dy}{dx} = \frac{wL^3}{24} = \frac{WL^2}{24}$$

therefore

$$\frac{dy}{dx} = \frac{WL^2}{24EI}$$

For $x = L$

$$EI\frac{dy}{dx} = -\frac{wL^3}{4} + \frac{wL^3}{6} + \frac{wL^3}{24} = -\frac{wL^3}{24} = -\frac{WL^2}{24}$$

therefore

$$\frac{dy}{dx} = -\frac{WL^2}{24EI}$$

Example   For a simply supported beam of length L carrying a concentrated load W at mid-span the variation of bending moment, M, with distance x from one end of the beam is given by the following expressions

(a) Between $x = 0$ and $x = L/2$

$$M = \frac{Wx}{2}$$

therefore

$$EI\frac{d^2y}{dx^2} = -M = -\frac{Wx}{2}$$

Integrating

$$EI\frac{dy}{dx} = -\frac{Wx^2}{4} + A$$

At mid-span $x = L/2$ and dy/dx (slope) $= 0$ therefore

$$A = \frac{WL^2}{16}$$

The maximum slope occurs at the support where $x = 0$, thus

$$EI\frac{dy}{dx} = \frac{WL^2}{16}$$

(b) Between $x = L/2$ and $x = L$

$$M = \frac{WL}{2} - \frac{Wx}{2}$$

therefore

$$EI\frac{d^2y}{dx^2} = -M = \frac{Wx}{2} - \frac{WL}{2}$$

Integrating

$$EI\frac{dy}{dx} = \frac{Wx^2}{4} - \frac{WLx}{2} + A$$

At mid-span $x = L/2$ and dy/dx (slope) $= 0$ therefore

$$A = \frac{WL^2}{4} - \frac{WL^2}{16} = \frac{3WL^2}{16}$$

$$EI \frac{dy}{dx} = \frac{Wx^2}{4} - \frac{WLx}{2} + \frac{3WL^2}{16}$$

Integrating again

$$EIy = \frac{Wx^3}{12} - \frac{WLx^2}{4} + \frac{3WL^2x}{16} + B$$

At a support, when $x = L$, y (deflection) $= 0$ therefore

$$B = \frac{WL^3}{4} - \frac{WL^3}{12} - \frac{3WL^3}{16} = -\frac{WL^3}{48}$$

$$EIy = \frac{Wx^3}{12} - \frac{WLx^2}{4} + \frac{3WL^2x}{16} - \frac{WL^3}{48}$$

The maximum deflection occurs at mid-span ($x = L/2$); substituting this

$$EIy_{max} = \frac{WL^3}{96} - \frac{WL^3}{16} + \frac{3WL^3}{32} - \frac{WL^3}{48} = \frac{WL^3}{48}$$

therefore

$$y_{max} = \frac{WL^3}{48EI}$$

The maximum slope occurs at a support when $x = L$, when

$$EI \frac{dy}{dx} = \frac{WL^2}{4} - \frac{WL^2}{2} + \frac{3WL^2}{16}$$

therefore

$$\frac{dy}{dx} = -\frac{WL^2}{16}$$

Example   For a cantilever of length L carrying a uniformly distributed load of w per unit length, the variation of bending moment, M, with distance x, measured from the free end, is given by

$$M = -\frac{wx^2}{2}$$

therefore

$$EI \frac{d^2y}{dx^2} = -M = \frac{wx^2}{2}$$

Integrating

$$EI \frac{dy}{dx} = \frac{wx^3}{6} + A$$

At the fixed end, when $x = L$, the slope $dy/dx = 0$ therefore

$$A = -\frac{wL^3}{6}$$

Integrating again

$$EIy = \frac{wx^4}{24} - \frac{wL^3x}{6} + B$$

At the fixed end, when $x = L$, the deflection $y = 0$ therefore

$$B = \frac{wL^4}{6} - \frac{wL^4}{24} = \frac{wL^4}{8}$$

The maximum deflection occurs at the free end ($x = 0$); substituting this

$$EIy_{max} = \frac{wL^4}{8} = \frac{WL^3}{8}$$

therefore

$$y_{max} = \frac{WL^3}{8EI}$$

The maximum slope is also at the free end; substituting $x = 0$

$$EI\frac{dy}{dx} = -\frac{wL^3}{6} = -\frac{WL^2}{6}$$

therefore

$$\frac{dy}{dx} = -\frac{WL^2}{6EI}$$

Example   For a cantilever with a concentrated load W at the free end the bending moment, M, expressed as a function of the distance x from the free end, is given by

$$M = -Wx$$

therefore

$$EI\frac{d^2y}{dx^2} = -M = Wx$$

Integrating

$$EI\frac{dy}{dx} = \frac{Wx^2}{2} + A$$

At the fixed end, when $x = L$, the slope $dy/dx = 0$ therefore

$$A = -\frac{WL^2}{2}$$

The maximum slope occurs at the free end, where $x = 0$, when

$$EI\frac{dy}{dx} = -\frac{WL^2}{2}$$

therefore

$$\frac{dy}{dx} = -\frac{WL^2}{2EI}$$

Integrating a second time

$$EIy = \frac{Wx^3}{6} - \frac{WL^2x}{2} + B$$

At the fixed end, when $x = L$, the deflection $y = 0$ therefore

$$B = \frac{WL^3}{2} - \frac{WL^3}{6} = \frac{WL^3}{3}$$

The maximum deflection occurs at the free end ($x = 0$), when

$$EIy_{max} = \frac{WL^3}{3}$$

therefore

$$y_{max} = \frac{WL^3}{3EI}$$

The determination of slope and deflection by means of the integration method illustrated above becomes cumbersome and laborious when applied to beams subject to more than one concentrated load. An integration method was devised by Macaulay to cover these more complex cases, but the analysis of this type of problem is beyond the scope of this book.

6.6 Deflection by Strain Energy

The strain energy, U, for a beam is given by

$$U = \int \frac{M^2}{2EI} \, dx$$

If a beam carries a single concentrated load W, the strain energy, U, must equal the work done when the load is slowly applied, and this is given by

$$U = \tfrac{1}{2} Wy$$

where y is the deflection produced. Therefore

$$y = \frac{2U}{W}$$

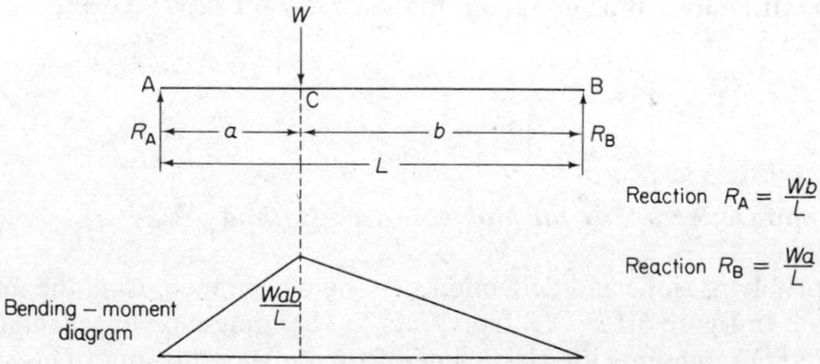

Reaction $R_A = \dfrac{Wb}{L}$

Reaction $R_B = \dfrac{Wa}{L}$

Bending – moment diagram

FIGURE 6.9

This method is only applicable to cases of beams carrying one concentrated load.

Consider a simply supported beam of length L carrying a single concentrated load W at a distance of a from one support (figure 6.9).

Considering the section of the beam between A and C, the variation of bending moment, M, with distance x from support A is given by

$$M = \frac{Wbx}{L}$$

The strain energy, U_1, for the section AC is given by

$$U_1 = \int_0^a \frac{M^2}{2EI} \, dx = \int_0^a \frac{W^2 b^2 x^2}{2EIL^2} \, dx$$

$$= \left[\frac{W^2 b^2 x^3}{6EIL^2} \right]_0^a$$

$$= \frac{W^2 b^2 a^3}{6EIL^2}$$

Similarly, it can be proved that the strain energy, U_2, for the section of the beam BC is given by

$$U_2 = \frac{W^2 a^2 b^3}{6EIL^2}$$

The total strain energy $U = U_1 + U_2$, so

$$U = \left(\frac{W^2 a^2 b^2}{6EIL^2} \right) (a + b)$$

$$= \frac{W^2 a^2 b^2}{6EIL}$$

But deflection $y = 2U/W$ therefore

$$y = \frac{Wa^2 b^2}{3EIL}$$

For a beam with a load W at mid-span, that is, $a = b = L/2$ we have

$$y = \frac{WL^3}{48EI}$$

6.7 Slope and Deflection by the Moment–Area Method

For some problems slope and deflection may be determined using the moment–area method. In figure 6.10 XYZ represents the bending-moment diagram for a beam, while PQ represents the elastic line of the deflected beam. CD is some chosen line. The tangents at P and Q make an intercept of length y on the chosen

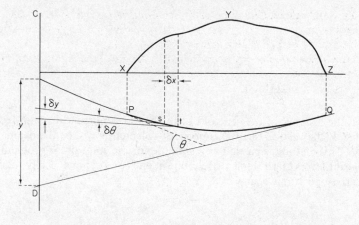

FIGURE 6.10 *Moment–area method for beam deflexion*

line. Consider a small length of the beam, δx. The bending moment, M, can be assumed to have a constant value over this short length δx. The change in slope over this small length is $\delta\theta$, where

$$\delta\theta = \frac{\delta x}{R}$$

where R is the radius of curvature of length δx. But

$$\frac{1}{R} = \frac{M}{EI}$$

therefore

$$\delta\theta = \frac{M\delta x}{EI}$$

therefore

$$\theta = \int_{X}^{Z} \frac{M}{EI}\, dx$$

But $\int_{X}^{Z} M\, dx = A$, where A is the area of the bending-moment diagram XYZ, therefore

$$\theta = \frac{A}{EI}$$

This is an important relationship since it means that the change in slope between any two points on a loaded beam is given by the area of the moment diagram between those points, divided by the flexural rigidity, EI. The intercepts of the tangents at s and t on the elastic line with the chosen line CD is δy, and since the slope is small, δy approximates to $x\delta\theta$. Therefore

$$\delta y = \frac{Mx\delta x}{EI}$$

and

$$y = \int_x^z \frac{Mxdx}{EI}$$

$$= \frac{A\overline{x}}{EI}$$

where $\overline{x}$ is the distance of the centroid of the bending-moment diagram from the chosen line. If the vertical line CD is properly selected, y will be the beam deflection. This method is particularly suitable for the solution of problems involving cantilevers (zero slope at the built-in end) and symmetrically loaded beams (zero slope at mid-span).

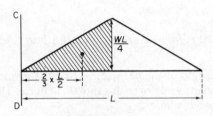

FIGURE 6.11

Example   To determine the slope and deflection for a simply supported beam of length L carrying a concentrated load of W at mid-span consider one-half of the bending-moment diagram (figure 6.11), and take the chosen line CD at one support. The area of half the bending-moment diagram is given by

$$A = \frac{1}{2}\frac{L}{2}\frac{WL}{4} = \frac{WL^2}{16}$$

The distance of the centroid of this area from the chosen line CD is

$$\overline{x} = \frac{L}{2}\frac{2}{3} = \frac{L}{3}$$

The maximum slope at the support A is given by

$$\frac{A}{EI} = \frac{WL^2}{16EI}$$

The maximum deflection at mid-span is given by

$$\frac{A\overline{x}}{EI} = \frac{WL^3}{48EI}$$

The elastic line is horizontal at mid-span, that is, zero slope.

Example   For a simply supported beam of length L carrying a uniformly distributed load of w per unit length consider one-half of the bending-moment diagram (figure 6.12), and take the chosen line CD passing through one support.

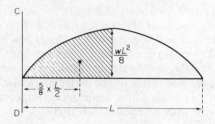

FIGURE 6.12

The area of half of the bending moment diagram is given by

$$A = \frac{2}{3}\frac{L}{2}\frac{wL^2}{8} = \frac{wL^3}{24}$$

The distance of the centroid of this area from the chosen line CD is

$$\bar{x} = \frac{L}{2}\frac{5}{8} = \frac{5L}{16}$$

The maximum slope at a support is given by

$$\frac{A}{EI} = \frac{wL^3}{24EI} = \frac{WL^2}{24EI}$$

where W (the total load) = wL. The maximum deflection at mid-span is given by

$$\frac{A\bar{x}}{EI} = \frac{5wL^4}{384EI} = \frac{5WL^3}{384EI}$$

6.8 Deflection by a Graphical Method

In section 5.4 a graphical method for the determination of shearing-force and beinding-moment diagrams was described. A similar method may be used to determine the elastic line of a deflected beam from the bending-moment diagram. The procedure is as follows. Divide the bending-moment diagram into a number of

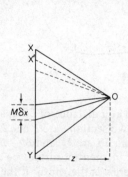
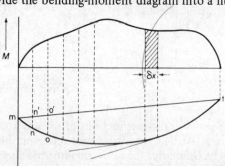

narrow strips of width δx. Draw a vertical line XY to the left of a pole O (figure 6.13). Mark off, to scale, the areas $M\delta x$ on the line XY. Join the marked points on the line XY to the pole O. The polygon mno . . . t may now be drawn by drawing a line mn parallel to XO, line no parallel to X'O, and so on.

After completing the polygon, the straight lines may be smoothed out into a curve, this curve being the elastic line of the deflected beam. The vertical lines nn', oo', etc., represent the magnitude of the beam deflection corresponding to those points. If the diagrams are drawn to scale, such that

$$1 \text{ mm} = g \text{ m of beam length}$$
$$1 \text{ mm on XY} = h \text{ m}^2 \ (M\delta x)$$
the distance from the pole O to the line XY is z mm

the 1 mm length of vertical ordinate in the elastic line curve will be equivalent to $ghz \ 10^{-3}/EI$ mm beam deflection at that point.

The polygon mno . . . t may be sheared through some angle to render the line mt horizontal. If this is done, the curve becomes a true representation of the elastic line for a horizontal simply supported beam.

6.9 Shear Stresses in Beams

In preceding sections we discussed the establishment of a direct stress within the material of a beam due to the action of a bending moment. One of the assumptions made in developing a theory of pure bending (section 6.2) was that there was no shear. A beam, however, is subject to a shear force and this will induce a shear stress on transverse sections of the beam.

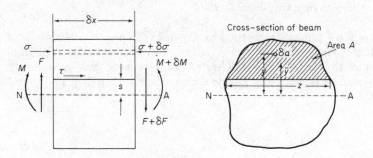

FIGURE 6.14

Consider two transverse sections of a beam situated at a distance of δx apart (see figure 6.14). The shearing forces acting on these two sections are F and $F + \delta F$ respectively. Similarly, the corresponding bending moments are M and $M + \delta M$. Let the shear stress developed at some distance s from the neutral surface be τ. This shear stress τ is acting on a plane of area $z\delta x$, where z is the width of the beam cross-section at that point.

The beam as a whole is in equilibrium, and therefore the shear stress τ acting on the plane of area $z\delta x$ must balance the difference of the longitudinal forces acting on the area A. This area, A, is the area of that part of the total section which is cut off by a plane parallel to the neutral surface at distance s from the neutral surface.

If the direct longitudinal stresses acting on an element of area δa on the two transverse planes are σ and $\sigma + \delta\sigma$, the difference of the longitudinal forces will be $\delta\sigma\,\delta a$. The difference in the longitudinal forces over the full area A of the cut-off section will be

$$\int d\sigma\,da$$

But the beam is in equilibrium, so

$$\int d\sigma\,da = \tau z\delta x$$

From the bending formula $\sigma = My/I$ therefore

$$\sigma + \delta\sigma = \left(\frac{M + \delta M}{I}\right)y$$

$$\delta\sigma = \frac{\delta My}{I}$$

$$\tau z\delta x = \frac{\delta M}{I}\int y\,da$$

$$\tau = \frac{\delta M}{\delta x\,Iz}\int y\,da$$

But $\int y\,da = A\bar{y}$, where A is the total area of the cut-off section and $\bar{y}$ is the distance of the centroid of this area from the neutral axis. Also the shearing force $F = dM/dx$ (from section 5.3), therefore

$$\tau = \frac{FA\bar{y}}{Iz}$$

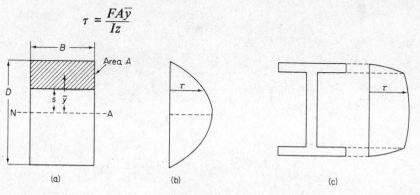

(a) (b) (c)

FIGURE 6.15

Example   For a rectangular section of width B and depth D the value of shear stress at distance s from the neutral axis is given by $FA\bar{y}/IB$ (see figure 6.15). The value of I for the section, about the neutral axis is

$$I = \frac{BD^3}{12}$$

Area A is given by $B[(D/2) - s]$, and distance $\bar{y}$ is $\frac{1}{2}[(D/2) + s]$ therefore

$$\tau_s = \frac{FB\left(\frac{D}{2} + s\right)\left(\frac{D}{2} - s\right)}{2B\frac{BD^3}{12}}$$

$$= \frac{6F}{BD^3}\left(\frac{D^2}{4} - s^2\right)$$

This relationship shows that the variation of shear stress, τ, with distance s from the neutral axis is of a parabolic form. The value of shear stress varies from zero at the upper and lower surfaces of the beam, when $s = D/2$, to a maximum of $\frac{3}{2}$ F/BD at the neutral axis, when $s = 0$.

The variation of the value of shear stress across a section of an I-beam is shown in figure 6.15c.

Summary

Second moment of area I
Bending formula

$$\frac{M}{I} = \frac{E}{R} = \frac{\sigma}{y}$$

TABLE 6.2

Section	Axis	I
	Diameter	$\frac{\pi D^4}{64}$
	XX	$\frac{BD^3}{12}$
	YY	$\frac{DB^3}{12}$
	XX	$\frac{BD^3 - bd^3}{12}$
	YY	$\frac{DB^3 - db^3}{12}$
	XX	$\frac{BD^3 - bd^3}{12}$

TABLE 6.3

Type of beam	Maximum deflection	Maximum slope
W cantilever, point load at free end, length L	$\dfrac{WL^3}{3EI}$ at free end	$-\dfrac{WL^2}{2EI}$ at free end
$W = wL$ cantilever, UDL, length L	$\dfrac{WL^3}{8EI}$ at free end	$-\dfrac{WL^2}{6EI}$ at free end
W simply supported, point load at mid-span, $\frac{L}{2}$ each side	$\dfrac{WL^3}{48EI}$ at mid-span	$\dfrac{WL^2}{16EI}$ at supports
$W = wL$ simply supported, UDL, length L	$\dfrac{5WL^3}{384EI}$ at mid-span	$\dfrac{WL^2}{24EI}$ at supports

Questions

6.1 Two beams, one made of PVC and the other of cast iron, have a length of 0.9 m and are each simply supported at their ends. Each beam has a rectangular section 25 mm wide and 18 mm deep and carries a single concentrated load of 270 N at a point 0.3 m from the left-hand end.

(a) Sketch neatly the shearing-force and bending-moment diagrams.
(b) Calculate the maximum tensile and compressive stresses in the material of each beam.
(c) In what way and for what reasons does the maximum deflection differ for the two beams?
(d) Describe briefly the effect of increasing the temperature slowly from 15°C to 200°C on each of the two beams.

(Ans. (a) M_{max} = 54 Nm; (b) 40 MN/m²)

(AEB)

6.2 A straight uniform horizontal beam is 40 cm long. The beam is of I-section, with a flange width of 20 mm and a total depth of 24 mm. The thickness of both upper and lower flanges is 5 mm and the thickness of the web is 8 mm. The beam is simply supported at each end and carries a uniformly distributed load of 2 kN/m over the complete length. If the bending stress must nowhere exceed 100 MN/m² what additional concentrated load may be tolerated at the centre of the span?

(Ans. 810 N)

(Brighton Polytechnic)

6.3   A horizontal girder is 6.5 m long and of I-section, 250 mm deep. It is supported at each end and carries a load of 50 kN at a point 2.5 m from one end. Calculate the maximum direct stress in the material of the beam and sketch the stress distribution over the section. ($I = 88.2 \times 10^6$ mm^4)
(*Ans.* 109 MN/m^2)

6.4   Calculate the second moment of area of an I-beam of the following dimensions: width of top and bottom flanges = 125 mm, thickness of each flange = 20 mm, total depth of beam = 150 mm, thickness of web = 15 mm.
    A cantilever of this section projects horizontally 5 m from a wall. Calculate the maximum tensile stress in the material due to its own weight. Calculate the greatest concentrated load which may be carried at the free end if the maximum stress is not to exceed 70 MN/m^2. (Density of beam material is 7.9×10^3 kg/m^3.)
(*Ans.* 23×10^{-6} m^4; 21.1 MN/m^2; 3.01 kN)

6.5   A beam is simply supported at its ends over a 5 m span and carries a concentrated load of 140 kN at a point 2 m from one end. The beam is 250 mm deep and the maximum stress in the material of the beam is 70 MN/m^2. What is the maximum deflection of the beam? ($E = 210 \times 10^9$ N/m^2)
(*Ans.* 4.6 mm)

6.6   Samples of glass, of dimensions 120 mm × 12 mm × 3 mm, are tested in the manner of a simply supported beam with a force applied at mid-span. The testing machine beam supports are set 100 mm apart, and the force required to fracture the samples is 45 N. Determine the tensile strength of the glass.
(*Ans.* 62.1 N/mm^2)

7

Theory of Torsion

7.1 Torsion of a Cylinder

If a material is subjected to twisting by the application of a couple a shear stress
will be induced within the material. If a couple is applied to a cylindrical rod in
such a way that the axis of the couple is coincident with the axis of the rod, then
the rod is said to be subject to pure torsion. At any point within the cross-section
of a rod subjected to pure torsion, there will be a pure shear stress established in a
direction normal to the radius of the rod at that point.

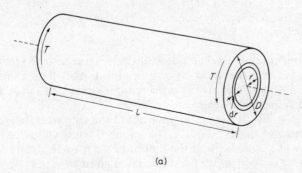

(a)

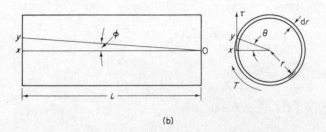

(b)

FIGURE 7.1   *Torsion of a cylinder*

For a cylinder of diameter D and length l (see figure 7.1a), consider first a small annular element of mean radius r and thickness dr. The cylinder is subjected to a torque T, and this causes a circumferential shear stress τ in the wall of the small element under consideration. If the torque is such that the left-hand end of this small element is twisted through an angle θ in relation to the right-hand end, a longitudinal line Ox on the surface of the element twists to position Oy (see figure 7.1b). For small angles of twist the shear strain γ that is developed is given by

$$\gamma = \frac{xy}{l} = \tan \phi$$

but the arc $xy = r\theta$, therefore

$$\gamma = \frac{r\theta}{l}$$

Within the elastic limit, the ratio of stress to strain is constant.

$$\frac{\tau}{\gamma} = G$$

where G is the modulus of rigidity. Substituting for γ, we have

$$\frac{\tau l}{r\theta} = G$$

This may be rewritten as

$$\frac{\tau}{r} = \frac{G\theta}{l} \tag{7.1}$$

For this small element, if the thickness dr is small, it can be assumed that the shear stress is constant across the thickness of the element. Also, if dr is small, the area of the section on which the shear stress τ acts approximates to $2\pi r\, dr$. Hence, the total shear force acting on this element is $\tau\, 2\pi r\, dr$.

The torque acting on this element is the moment of the tangential shear force about the longitudinal axis, and is $\tau\, 2\pi r^2\, dr$. The torque T acting on the complete solid shaft is the sum of the moments of the tangential shear forces acting on all the small elements that go to make up the shaft, and is given by

$$T = \int_0^{D/2} \tau\, 2\pi r^2\, dr$$

But, from equation 7.1 $\tau/r = G\theta/l$. Substituting for τ, we have

$$T = \frac{G\theta}{l} \int_0^{D/2} 2\pi r^3\, dr$$

$$= \frac{G\theta}{l} \frac{\pi D^4}{32}$$

From section 6.1 for a circular section

$$\frac{\pi D^4}{32} = J$$

where J is the polar moment of area. Therefore

$$T = \frac{G\theta}{l} J$$

or

$$\frac{T}{J} = \frac{G\theta}{l} \tag{7.2}$$

Combining equations 7.1 and 7.2 we have

$$\frac{T}{J} = \frac{G\theta}{l} = \frac{\tau}{r} \tag{7.3}$$

This is the general relationship for the pure torsion of cylindrical bodies. For the correct application of this formula the units used must be consistent; for example

torque	T	Nm
polar moment of area	J	m^4
modulus of rigidity	G	N/m^2
angle of twist	θ	rad
length	l	m
shear stress	τ	N/m^2
radius	r	m

From equation 7.3 it will be seen that the magnitude of the shear stress is not constant throughout the section but varies directly with the radial distance from the axis of the shaft.

The relationships in equation 7.3 are also applicable to hollow cylindrical shafts of uniform wall-thickness. The polar moment of area, J, for a hollow cylindrical shaft with an external diameter D and an internal diameter d is given by

$$J = \frac{\pi}{32}(D^4 - d^4)$$

The general torsion formula is valid, provided that the following assumptions are satisfied

(1) the applied torque is pure and is acting about the longitudinal axis of the shaft,
(2) the applied torque is of constant magnitude at all points along the shaft,
(3) the shaft is of uniform cross-section at all points along its length,
(4) the shaft is straight,
(5) the stresses are within the elastic limit of the material, and
(6) a plane cross-section before twisting remains plane after twisting.

The torsional stiffness of a shaft is the torque per radian of twist, T/θ. From equation 7.2, the torsional stiffness, $T/\theta = GJ/l$.

Example   A solid circular shaft of diameter 75 mm is subjected to torsion causing a twist of $1°$ $15'$ per metre. Calculate the maximum shear stress in the shaft and the torque transmitted. Assume the modulus of rigidity of the shaft material is $80\ GN/m^2$.

$$\frac{\tau}{r} = \frac{G\theta}{l}$$

$$\tau = \frac{G\theta r}{l}$$

The angle of twist is $1° \; 15' = (1.25 \times 2\pi)/360$ rad/m

$$\tau = \frac{80 \times 10^9 \times 1.25 \times 2\pi \times 75 \times 10^{-3}}{360 \times 2}$$

$$= 65.45 \times 10^6 \; N/m^2$$

The torque transmitted is given by

$$T = \frac{G\theta J}{l}$$

$$J = \frac{\pi D^4}{32}$$

where $D = 75$ mm (0.075 m)

$$T = \frac{80 \times 10^9 \times 1.25 \times 2\pi \times \pi (0.075)^4}{360 \times 32}$$

$$= 54.23 \times 10^3 \; Nm$$

Example (a) Calculate the maximum torque that a hollow cylindrical shaft of external diameter 200 mm and internal diameter 100 mm can carry if the maximum shear stress is not to exceed 60 MN/m². (b) What would be the minimum diameter of a solid shaft capable of carrying the same torque, if also subject to a maximum shear stress of 60 MN/m²?

(a) The maximum shear stress occurs at the periphery of the shaft ($r = 0.1$ m).

$$T = \frac{\tau J}{r}$$

$$J = \frac{\pi (D^4 - d^4)}{32} = \frac{\pi (0.2^4 - 0.1^4)}{32}$$

$$= \frac{\pi 0.0015}{32} \; m^4$$

$$T = \frac{60 \times 10^6 \times \pi \times 0.0015}{0.1 \times 32}$$

$$= 88.4 \times 10^3 \; Nm$$

(b) For a solid shaft

$$r = \frac{D}{2} = \frac{\tau J}{T}$$

$$= \frac{\tau \pi D^4}{32T}$$

$$D = \sqrt[3]{\left(\frac{16T}{\tau \pi}\right)}$$

$$= \sqrt[3]{\left(\frac{16 \times 88.4 \times 10^3}{60 \times 10^6 \times \pi}\right)}$$

$$= 0.196 \text{ m (196 mm)}$$

7.2 Transmission of Power

Rotating shafts are used for the transmission of power. The power, P, transmitted by a rotating shaft is given by

$$P = T2\pi n$$

where P is the power transmitted (W), T is the torque (Nm), and n is the rotational speed of the shaft (rev/s).

Example   Determine the diameter of a solid shaft that will transmit 500 kW at 240 rev/min. The maximum shear stress in the material of the shaft is not to exceed 40 MN/m².

When transmitting 500 kW power at 240 rev/min the shaft will be subject to a torque T given by

$$T = \frac{P}{2\pi n} = \frac{500 \times 10^3 \times 60}{2\pi \times 240}$$

$$= 19.9 \times 10^3 \text{ Nm}$$

From the general torsion equation

$$\frac{T}{J} = \frac{\tau}{r}$$

$$\frac{T}{\tau} = \frac{\pi D^4}{32 \times \frac{D}{2}} = \frac{\pi D^3}{16}$$

$$D = \sqrt[3]{\left(\frac{16T}{\tau \pi}\right)}$$

$$= \sqrt[3]{\left(\frac{16 \times 19.9 \times 10^3}{40 \times 10^6 \times \pi}\right)}$$

$$= 0.136 \text{ m (136 mm)}$$

7.3 Energy Stored in a Shaft

When a torque is gradually applied to a shaft to some value T, resulting in an angle of twist θ, the strain energy, U, stored in the shaft is given by

$$U = \frac{1}{2} T\theta \quad \text{(assuming that all strain is elastic)}$$

$$= \frac{1}{2} \times \frac{\tau J}{r} \times \frac{\tau l}{rG}$$

$$= \frac{\tau^2}{2G} \times \frac{Jl}{r^2}$$

For a solid shaft

$$J = \frac{\pi D^4}{32} = \frac{\pi r^4}{2}$$

therefore

$$U = \frac{\tau^2}{2G} \times \frac{\pi r^2 l}{2}$$

But $\pi r^2 l$ = the volume of the shaft, therefore

$$U = \frac{\tau^2}{4G} \times \text{ volume of shaft}$$

Similarly, for a hollow shaft where $J = (\pi/32)(D^4 - d^4)$, it can be shown that the total strain energy U is given by

$$U = \frac{\tau^2}{4G} \times \frac{D^2 + d^2}{D^2} \times \text{ volume of shaft}$$

Summary

Torsion equation for circular shafts

$$\frac{T}{J} = \frac{G\theta}{l} = \frac{\tau}{r}$$

Polar moment of area J: for solid circular sections

$$J = \frac{\pi D^4}{32}$$

For hollow circular sections

$$J = \frac{\pi}{32}(D^4 - d^4)$$

Strain energy $U = \frac{1}{2} T\theta$

for solid circular shafts

$$U = \frac{\tau^2}{4G} \times \text{ volume of shaft}$$

for hollow shafts

$$U = \frac{\tau^2}{4G} \times \frac{D^2 + d^2}{D^2} \times \text{ volume of shaft}$$

Power transmitted by a rotating shaft

$$P = T2\pi n$$

Questions

7.1 A bronze shaft of circular cross-section is subjected to a torque of 1600 Nm. The maximum shear stress is not to exceed 62×10^6 N/m². Calculate (a) the diameter of the shaft, (b) the angle of twist in degrees per m length. Give the main constituents of a bronze suitable for this shaft.
Modulus of rigidity for bronze is 46×10^9 N/m².
(*Ans.* (a) 50.8 mm; (b) 3.04°)

(AEB)

7.2 A hollow turbine shaft is to transmit power at 240 rev/min. If the shaft is 500 mm external diameter with 20 mm wall thickness and the maximum shear stress permitted is 70 MN/m², find the maximum power it can transmit. What is the diameter of the equivalent solid shaft and what is the percentage saving in weight of the hollow shaft over the solid shaft?
(*Ans.* 12.25 MW; 329 mm; 64.9%)

(Derby College of Technology)

7.3 A hollow steel shaft of circular cross-section is to have an outside diameter of 35 mm. The shear stress must not exceed 60 MN/m². Using the formula $\tau/r = T/J = G\theta/l$ determine the necessary inside diameter if the shaft must transmit 20 kW at a speed of 1000 rev/min. If the complete length of the shaft twists through an angle of 0.75° while transmitting this power, how much elastic energy is stored in it?
Assume $G = 80$ GN/m².
(*Ans.* 22.8 mm; 1.885 J)

(Brighton Polytechnic)

7.4 A solid propeller shaft is 150 mm in diameter and the material has a safe working stress in shear of 60 MN/m². Calculate the maximum power that may be transmitted at 165 rev/min.
A tubular shaft of the same material and of 200 mm external diameter is required to transmit the same power at the same speed. Calculate the necessary internal diameter and the percentage saving in weight, as compared with the solid shaft.
(*Ans.* 688.2 kW; 174.5 mm; 57.3%)

7.5 A hollow steel shaft with an outside diameter of 100 mm and an inside diameter of 75 mm is to transmit 75 kW at 160 rev/min. Calculate the maximum shear stress in this shaft.

Calculate the diameter of a solid shaft of the same material that will transmit the same power for the same value of maximum stress. Determine the ratio of angle of twist per metre length for the two shafts.
(*Ans.* 33.33 MN/m^2; 88 mm; 0.88.)

7.6   It is required that a bronze shaft transmit a maximum power of 1000 kW at a speed of 200 rev/min. Determine (a) the smallest-diameter solid shaft which may be used, if the maximum shear stress in the shaft is not to exceed 45×10^6 N/m^2; (b) the twist per unit length in the shaft when it is transmitting maximum power. Assume modulus of rigidity for bronze is 45×10^9 N/m^2.
(*Ans.* (a) 0.175 m; (b) 0.0114 rad or 0° 39′)

(AEB)

8

Plastic Behaviour

8.1 Plastic Flow in Metals

Many materials possess an elastic limit and when subject to stress they strain in an elastic manner up to a certain point. Beyond this point, known as the elastic limit, the strain developed is no longer directly proportional to the applied stress, and also, the strain developed is no longer fully recoverable. If the stress is removed elastic strain is recovered but the material will be left in a state of permanent, or plastic, strain. The mechanism of plastic deformation is not the same for all classes of materials and it is necessary to consider the various materials groups separately.

Metals in general are characterised by possessing high elastic modulus values, and also the ability to be strained in a plastic manner. Some metals will begin to deform plastically at very low values of stress and will yield to a very considerable extent before fracture occurs. Such metals are termed *ductile*. Other metals and alloys show little plastic yielding before fracture. These latter materials are termed brittle. Plastic deformation in metals may take place by the process of slip, or by twinning. Slip is the more common deformation process encountered and will be dealt with first.

An early theory evolved to explain plastic deformation in metals was the block-slip theory. In this it was postulated that when the yield stress of the metal was exceeded plastic deformation took place by the movement of large adjacent blocks of atoms sliding across certain planes — *slip planes* — within the crystal (see figure 8.1). The block-slip theory accounted for many of the observed phenomena but it also possessed several deficiencies. Two of the drawbacks to this theory were

(1) plastic deformation of metals commences at a stress level that is approximately 0.1 per cent of the value of stress calculated to be necessary to cause the movement of large blocks of atoms within a crystal;

(2) metals work harden as a result of plastic deformation, that is, as plastic flow takes place the level of stress necessary to bring about continued deformation is increased. If, however, the block-slip theory were correct, once the yield stress of the material is exceeded continued plastic strain should be sustained with no further increase in stress being necessary.

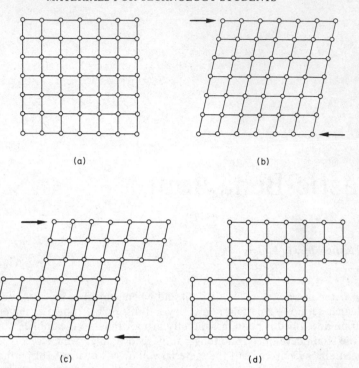

(a)   (b)

(c)   (d)

FIGURE 8.1 *Plastic deformation according to the block-slip theory;* (a) *original*
unstressed condition; (b) *elastic strain only, specimen will revert to*
(a) when the stress is removed; (c) *elastic and plastic deformation –*
when the stress is removed the specimen will appear as in (d); (d)
unstressed, no elastic strain but permanent plastic strain remains

Present-day theories of plastic flow in metals are based on the existence of
small imperfections, or defects, within crystals. These are structural defects, called
dislocations, and plastic deformation is due to the movement of dislocations across
the slip planes of a crystal under the action of an applied stress. In the development of the block-slip theory it was stated that the slip planes are generally the
most densely packed atomic planes within a crystal system. In the dislocation
theory of plastic deformation dislocations move across the most densely packed
planes of atoms. The calculated stress required to bring about the movement of
dislocations is of the same order of magnitude as observed yield stresses in metals.
In recent years it has been possible to produce small single crystals that are virtually
defect-free. These small perfect crystals, called *whiskers*, possess properties close
to the very high theoretical strengths predicted for perfectly crystalline metals.

8.2 Slip Planes

The crystal planes on which slip generally takes place are those that possess the
highest degree of atomic packing. The direction of slip within a slip plane is the
direction of greatest atomic line density. The majority of metals crystallise as

close-packed hexagonal, face-centred cubic, or body-centred cubic, and slip within these three systems will be discussed.

In the close-packed hexagonal system the planes with the greatest density of atomic packing are the basal planes (001) and within planes of this type there are three possible slip directions (see figure 8.2). Within any one hexagonal crystal space lattice there is only one series of parallel planes of this type.

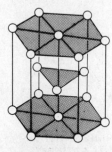

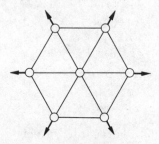

(a) Unit cell, showing basal
    slip planes

(b) Plan view of basal plane, showing
    the slip directions

FIGURE 8.2 *Slip planes and directions in the close-packed hexagonal system*

In the face-centred cubic system the most densely packed planes of atoms are those of the $\{111\}$ family. Within each (111) plane there are three possible slip directions, [1̄10] directions (see figure 8.3). There are four sets of planes of this type, each set occurring at different inclinations. Consequently, no matter from what direction relative to the crystal a direct force is applied there will be resolved shearing forces acting on several slip planes and at least one slip system will be inclined in such a way that plastic deformation can occur. This means that face-centred cubic crystals are comparatively soft and ductile.

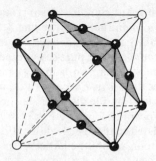

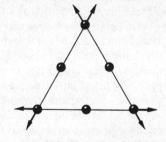

(a) Unit cell, showing a (111) slip plane

(b) Plan view of (111) plane, showing
    the slip directions

FIGURE 8.3 *Slip planes and directions in the face-centred cubic system*

In the body-centred cubic system the most densely packed planes are those of the {110} type. Within this type of plane there are two possible [111] slip directions (see figure 8.4).

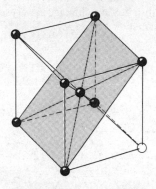 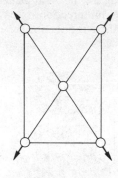

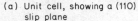

(a) Unit cell, showing a (110)   (b) Plan view of (110) plane showing
    slip plane   the slip directions

FIGURE 8.4   *Slip planes and directions in the body-centred cubic system*

Slip in body-centred cubic metals is more complex than in the other two systems already discussed. In addition to slip taking place on (110) planes, slip may also occur on planes of the (112) and (123) types. Although there are numerous possible slip directions within body-centred cubic crystals, metals crystallising in this form are generally harder and less ductile than face-centred cubic metals. This is thought to be due to the body-centred cubic lattice being less densely packed than the face-centred cubic lattice. An increase in temperature will bring more slip planes into action, and also reduce the value of critical shearing stress. This latter statement also applies to other forms of metallic crystal.

8.3 Role of Dislocations

Crystal space lattices are not perfectly crystalline but contain certain structural defects. These may be classified as either line or point defects. The line defects are known as *dislocations*, and these may be classified as either *edge*-type or *screw*-type. In practice, dislocation lines are rarely of the pure-edge or pure-screw type but are mixed dislocations, that is, lines of dislocation containing both an edge and a screw component. An important property of a dislocation is its *Burgers vector, b*, which indicates the extent of lattice displacement caused by the dislocation. The Burgers vector also indicates the direction in which slip will occur. Figure 8.5 is a representation of an edge dislocation. It can be considered that the edge dislocation is due to the presence of an additional half-row of atoms within the lattice. If an atom to atom circuit is described within a portion of regular lattice, as shown at the bottom left of figure 8.6a, it will be a complete closed circuit. If, however, a similar circuit is described around the dislocated portion of lattice the start and finish will not be coincident. The distance SF in figure 8.6a will be the Burgers vector, *b*. In an edge dislocation the Burgers vector is normal to the dislocation line.

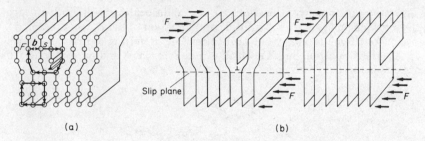

(a)   (b)

FIGURE 8.5 *Edge dislocation:* (a) *representation of a portion of crystal lattice*
containing an edge dislocation, dislocation shown as ⊥ and Burgers
*vector as **b**;* (b) *application of shear stress F causes the dislocation*
to move along the slip plane until it leaves this section of lattice
causing an increment of plastic deformation

When a shearing stress is applied to a section of crystal lattice, and this stress is
beyond the elastic limit, minor atomic movements will occur causing the dislocation line to move through the lattice (see figure 8.5b). The magnitude of the stress
necessary to initiate the movement of a dislocation acrosss a slip plane is considerably less than that which would be required to bring about block slip. It will be
seen that plastic flow occurs in the direction of the Burgers vector.

A diagrammatic representation of a screw dislocation is shown in figure 8.6. In
this case a Burgers circuit describes a helical (screw) path and the Burgers vector,
b, is in the same direction as the line of the dislocation. It should be noted, however, that plastic flow under the action of a shearing stress still occurs in the
direction of the Burgers vector.

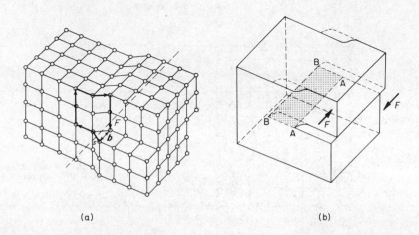

(a)   (b)

FIGURE 8.6 *Screw dislocation:* (a) *representation of a lattice containing a screw*
dislocation (b) *application of a shear stress F will cause the*
dislocation to move from AA to BB − the slipped area is shaded

As stated earlier many dislocations are mixed dislocations with both an edge and a screw component. The magnitude and direction of the Burgers vector are constant for all points on any one dislocation line. Referring to figure 8.7 it will be seen that the dislocation line XY which is pure screw at X and pure edge at Y has the same vector, *b*, at each end.

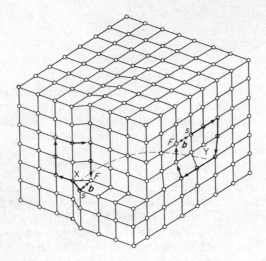

FIGURE 8.7 *Line of dislocation which is screw at* X *and edge at* Y

Crystals invariably contain numerous point defects in addition to the type of structural defect described above. The principal types of point defect are the *vacancy*, the *substitutional* defect and the *interstitial* defect (see figure 8.8). The

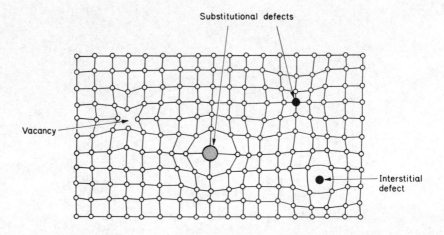

FIGURE 8.8 *Lattice containing several types of point defect; note that substitutional atoms may be larger or smaller than the parent atoms causing positive or negative lattice strain*

latter two types are due to the presence of 'stranger' atoms of other elements that are present either as impurities or are deliberately added as alloying elements. The interstitial type occurs when the stranger atoms are very small in comparison with the atoms of the parent metal, while the substitutional type occurs when parent and stranger atoms are comparable in size. It will be noticed from figure 8.8 that strain is developed within a crystal lattice in the neighbourhood of point defects. Some of the possible interactions between point and line defects and the influence of such defects upon the properties of metallic crystals will be discussed later in this chapter.

8.4 Polycrystalline Metals

In the foregoing chapters the mechanisms of plastic deformation within metallic crystals have been discussed. Metallic materials, however, are not normally in the form of single crystals, but are composed of many crystals, or *grains*, and in many cases these small crystals have a random orientation. For a polycrystalline sample of a close-packed hexagonal metal, such as zinc, it will be apparent that while the slip planes in some crystals might be favourably inclined for slip under the action of an applied stress, other crystals may not be aligned in a suitable direction. Plastic deformation of the favourably positioned crystals will be hindered, or even completely prevented, by unfavourably placed adjacent crystals (see figure 8.9). The crystal boundaries (grain boundaries) will also hinder plastic deformation. These boundaries are not simple planes but are transition zones between adjacent crystals of differing orientation (see figure 8.9). These transition zones, which may

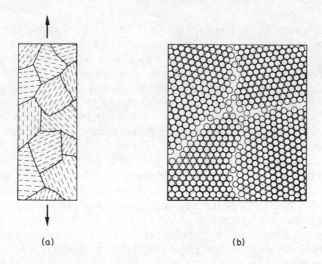

(a)   (b)

FIGURE 8.9   *Polycrystalline metals:* (a) *random orientation of slip planes (shown dotted) – some crystals are more favourably placed than others for plastic deformation;* (b) *representation of grains and boundaries*

be of several atoms in thickness, do not possess regular crystal planes and consequently act as barriers to the movement of dislocations. From this it follows that
the grain boundaries in a pure metal† tend to be stronger than the crystal grains
and that consequently a sample composed of a large number of very small crystals
will be stronger than another sample of the same metal containing a smaller number
of larger crystal grains. (This is not true for all temperatures. At elevated temperatures the grain boundaries become weaker than the crystals and plastic deformation
may occur through 'flow' of the grain-boundary material.)

A single crystal will exhibit anisotropy, that is, it will possess different properties
in different directions. This is very evident in the case of hexagonal metals that
possess only one series of parallel slip planes. However, a polycrystalline material
in which the orientation of the individual crystals is a purely random arrangement
will, on a macro scale, be virtually isotropic.

The deformation behaviour of the principal crystal types in polycrystalline
samples of pure metals may be summarised as follows.

(1) Close-packed hexagonal metals are generally brittle, or of low ductility
only. If ductility occurs, it usually only exists within a narrow range of
temperatures.
(2) Face-centred cubic metals are invariably highly ductile over a wide range
of temepratures.
(3) The body-centred cubic system is less favourable to ductility than the
face-centred cubic system. The body-centred cubic system contains both
ductile and brittle metals, but even the ductile metals in this system often
only show good ductility over a fairly small range of temperature.

8.5 Work Hardening and Recrystallisation

When a metal is stressed beyond its elastic limit, dislocations within the crystals
begin to move and plastic deformation begins. The dislocations cannot move freely
throughout the whole of the crystalline metal. Their progress will be impeded by
potential barriers such as interstitial and substitutional point defects, and grain
boundaries. An increased force will be necessary so that the dislocations can cross
the potential barriers and cause further increments of plastic deformation. In cubic
crystals, where slip may be occurring due to the simultaneous movement of several
dislocations on intersecting slip planes, the dislocations may interfere with one
another's movement.

In other words, the level of applied stress must be continually increased if
plastic deformation is to be continued, and the greater the amount of plastic
deformation that is given to the metal the larger will be the force necessary to
continue plastically deforming the material. The material becomes strengthened,
and this phenomenon is termed *work hardening* or *strain hardening*. This principle
is used in practice for metal strengthening. The strength of a ductile metal may be
increased considerably by plastically deforming it by means of a process such as
cold rolling or drawing (see table 8.1). Work hardening is accompanied by a decrease
in ductility and an increase in the electrical resistivity of the metal.

†In some alloys and impure metals there may be segregation of alloy constituents, or impurities,
to grain boundaries. In some cases this renders the crystal boundaries weak, for example,
copper is embrittled by the segregation of traces of bismuth impurity to the grain boundaries.

The visible effect on the microstructure of severely cold deforming a metal by a process such as rolling will be to cause the randomly orientated crystal grains to become broken up, strained, and aligned in the general direction of working. Each individual crystal of the original material will have been strained both elastically and plastically. When the deforming force is removed there should be recovery of all the elastic strain but in practice the complete elastic recovery of any individual crystal will be hindered by the rigidity of the surrounding crystals. This will lead to the presence of some locked-in elastic strain in the plastically deformed metal. Consequently the material will be in a state of some internal stress.

TABLE 8.1

Effect of cold rolling on the properties of commercial purity aluminium

Reduction in thickness (%)	Condition	Hardness (V.P.N.)	Tensile strength (MN/m^2)	Elongation (%)
0	Annealed	20	92	40
15	$\frac{1}{4}$ hard	28	107	15
30	$\frac{1}{2}$ hard	33	125	8
40	$\frac{3}{4}$ hard	38	140	5
60	Fully work-hardened	43	155	3

When a material in this condition is heated, changes will begin to take place. These changes may be classified under three headings (a) stress relief, (b) recrystallisation, and (c) grain growth.

As the temperature of the material is raised so the vibrational energies of the individual atoms are increased and atomic movements can occur. Comparatively minor atomic movements result in the removal of the residual stresses associated with the locked-in elastic strains. This change, which occurs at comparatively low temperatures, has a negligible effect on the strength and hardness of the material, and the microstructure of the metal is unchanged in its appearance. When the temperature is raised further the process of recrystallisation begins. New unstressed crystals begin to form and grow from nuclei until the whole of the material has a structure of unstressed polygonal crystals (see figure 8.10).

This change in structure is accompanied by a reduction in hardness, strength, and brittleness to the original values prior to plastic deformation. The temperature at which recrystallisation occurs is, for a pure metal, within the range from one-third to one-half of the melting temperature (K). The recrystallisation temperature is not, however, constant for any material since its value is affected by the amount of plastic deformation prior to heating. If the temperature is raised further, grain growth may occur following the completion of recrystallisation, with some crystal grains growing in size at the expense of others by a process of grain-boundary migration.

The industrial process of annealing† is a heat treatment allowing recrystallisation to take place with consequent softening of work-hardened materials. Some metals, such as lead, have recrystallisation temperatures below room temperature. This means that they cannot be work hardened at ordinary temperatures.

†The 'annealing' treatment for steels, while still being a softening treatment is based on a different principle.

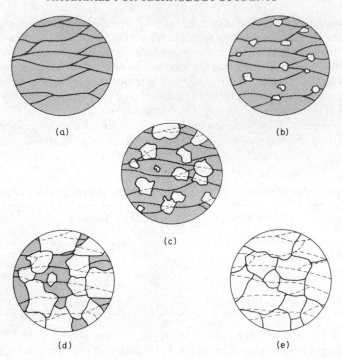

FIGURE 8.10   *Recrystallisation:* (a) *original cold-worked structure;* (b) *nuclei forming;* (c) *and* (d) *new grains growing from nuclei;* (e) *recrystallisation complete*

8.6 Solid Solution Strengthening

As stated earlier, the presence of interstitial and substitutional point defects introduces strain in the crystal lattice and causes a disturbance to the regularity of the crystal planes. This strain makes it more difficult for dislocations to be moved through the lattice. In other words, a greater stress will be required to cause plastic deformation in a metal containing this type of point defect than that needed in a pure metal free from such defects. The presence of these defects has the effect of strengthening a metal.

The alloying of metals to create substitutional or interstitial structures (termed solid solutions) is a valuable technique for increasing the strength of metals. The amount by which the strength of a metal is raised will depend on the total amount of strain developed in the lattice and this, in turn, is related to the amount of solute present and the magnitude of the difference between the atomic diameters of solvent and solute atoms.

Copper and nickel are both metals that crystallise in the face-centred cubic form. Alloys of these two elements always form a face-centred cubic solid solution, irrespective of the relative proportions of each element. The respective diameters of copper and nickel atoms differ by 2.5 per cent. Both the addition of nickel to copper and the addition of copper to nickel result in the development of strain in the lattice giving an increase in the strength and hardness of each pure metal. The properties of alloys of copper and nickel are given in table 8.2.

TABLE 8.2

Solid solution strengthening – properties of annealed copper–nickel solid solutions

| Composition | | | | | |
Cu (%)	Ni (%)	Hardness (V.P.N.)	Tensile strength (MN/m²)	Elongation (%)	Resistivity (Ωm)
100	–	60	210	65	1.67×10^{-8}
95	5	65	230	50	2.8×10^{-8}
90	10	70	250	45	5.1×10^{-8}
80	20	75	315	45	1.1×10^{-7}
70	30	80	370	45	2.8×10^{-7}
60	40	90	430	45	5.5×10^{-7}
30	70	120	520	45	3.8×10^{-7}
–	100	95	450	50	6.84×10^{-8}

The presence of stranger atoms in solution within the parent metal's crystal space lattice also affects the movement of valency electrons through the crystal, so increasing the electrical resistivity of the metal.

8.7 Dispersion Strengthening

A third principle for the strengthening of metallic materials is termed dispersion hardening. A very finely dispersed second phase distributed throughout the crystal lattice will provide a series of barriers to the passage of dislocations, and considerably higher stresses will be necessary to cause yielding and plastic deformation than would be the case for a single-phase material. This dispersed second phase may be metallic or non-metallic. An example of a non-metallic dispersion is the use of thorium oxide to increase the high-temperature strength of tungsten wire in electric-lamp filaments. Sintered aluminium powder (S.A.P.) is another example. The compacting and sintering of finely powdered aluminium produces aluminium containing a very fine dispersion of aluminium oxide.

Precipitation hardening (age hardening) is another dispersion-hardening process, but in this case the finely divided second phase is produced by giving an alloy a specific type of heat treatment. The types of alloy that may respond to this treatment are those in which partial solid solubility occurs (see sections 9.5 and 9.6).

8.8 Hardness and Ductility

Hardness is a property of metals that is often quoted, but hardness is not capable of definition without quoting the method used for the determination of hardness. One definition of hardness is the resistance of a material to abrasion. From this definition we can determine the relative hardness of a material by deciding if the material is abraded by, or abrades, a known standard. Mohs' scale of hardness is based on this principle. The scale consists of ten natural minerals ranging from soft talc (number 1 on Mohs' scale) to diamond (number 10 on Mohs' scale). Mohs' scale is suitable for relating the hardness of solids that are not capable of being deformed in a plastic manner, namely, minerals, ceramics and glasses.

The second definition of hardness is the resistance of a material to plastic deformation. Several tests have been devised to measure the resistance of a material to plastic deformation. Most of these rely on forcing a hard indenter of known geometry into the surface of the material under test for a fixed time under the action of a known force. After the force is removed either the area or the depth of the permanent impression made is measured. The Brinell, Vicker's diamond, and Rockwell tests for hardness are all based on the indentation principle.

If, for a plastically deformable solid, hardness is defined as the resistance to deformation then it follows that there should be some relationship between hardness and the strength of the solid. In general terms, the strength of a metal increases with increase in hardness, but there is no rule of general application relating hardness to strength. Several empirical relationships have been devised, each applicable to a particular group of materials. One that appears to hold fairly well for most steels, other than heavily cold-worked steels, or austenitic steels is

$$\text{tensile strength (MN/m}^2) = 3.4 \times \text{hardness (Vicker's diamond)}$$

Ductility and brittleness are other terms used to describe some of the properties of materials. These terms cannot be precisely defined, but they are all interconnected. Ductility refers to the ability of a material to be deformed plastically, in particular its ability to be drawn through a die to a smaller cross-section. In general, the ductility of a metal is reduced as the hardness increases, although there are a few notable exceptions to this trend. A ductile fracture is failure preceeded by a considerable amount of plastic deformation in the material.

The measurable quantities that are generally taken as a guide to the ductility of a metal are the values of percentage elongation and percentage reduction in area that can be obtained in a tensile test to destruction.

Brittleness is the opposite of ductility. A material is classed as brittle when the fracture is preceeded by little or no plastic deformation. Some metals, some polymer materials, and all ceramics and glasses show brittle behaviour.

Metals that crystallise as face-centred cubic are generally ductile over a very wide range of temperature. Hexagonal-structured metals are usually brittle at ordinary temperatures but may possess ductile behaviour over a small range of elevated temperatures. Some body-centred cubic metals are ductile at ordinary temperatures while others are brittle. Body-centred cubic metals, like those crystallising as hexagonal, undergo a change from brittle to ductile behaviour as the temperature is raised.

8.9 Behaviour of Ceramics under Stress

Ceramics are also crystalline materials, but there are considerable differences between the properties of ceramics and those of metals. Ceramics are generally brittle materials that fracture without yielding in a plastic manner. Yet if dislocations exist in metallic crystals it would be logical for similar imperfections to occur in ceramic crystals. Dislocations and other imperfections are found in non-metallic crystalline materials, and plastic flow under the action of a stress can occur in specially prepared crystals of some ceramics. Investigation has shown that such deformation is due to the movement of dislocations. However, because ceramics have a different type of interatomic bonding and possess different structures from ductile metals, the behaviour of the dislocations is different.

Although it is interesting to note that plastic flow can occur in non-metallic crystals under certain circumstances, polycrystalline ceramics are invariably non-ductile and fail in a brittle manner. The fracture strength in tension of a polycry-stalline ceramic is normally less than that of a single crystal of the same material and this is probably due to the formation of small cracks at grain boundaries. The compressive strengths of ceramics are very high and are considerably greater than their strengths in tension. A small crack will act as a point of stress concentration, and when stressed in tension the applied stress will cause small cracks to open up and grow in length leading to early fracture. A further influence on the strength and brittle nature of ceramic materials is porosity. By the very nature of the manufacturing processes involved many ceramics are of less than theoretical density due to porosity. An increase in the porosity of a material will lead to a decrease in the fracture strength.

Plastic flow in clays is not due to dislocation movement, but is an entirely different phenomenon. The clay minerals, which are complex hydrated alumino-silicates, form layer-type lattices giving rise to extremely thin plate-like crystals (see section 3.7). Owing to the electron distribution within the crystal the thin plate crystal is polarised giving rise to the existence of van der Waals forces between adjacent plates. With strong interatomic bonding, but relatively weak inter-crystal bonds, adjacent plate crystals may slip relative to one another. The plasticity of the clay will be considerably increased by the adsorption of water. The water molecule is also polarised and will be readily adsorbed on the crystals giving water-lubricated surfaces. Clay plasticity is completely removed by 'firing' the clay at some temperature in the range $1000°C$ to $1400°C$. During the firing operation the adsorbed water is first driven off, and then the water of crystallisation is lost altering the crystal structure of the material, and setting up strong interatomic bonds between the original plate-like crystals.

8.10 Glasses

Inorganic glasses, as mentioned in section 3.8, are complex silicates in an amorphous condition. At elevated temperatures glasses behave as viscous fluids and deformation under stress is time dependent. At lower temperatures glasses obey Hooke's law, are brittle, and fracture without plastic deformation. Like ceramics, glasses are considerably stronger in compression than in tension. The theoretical tensile strength of a glass is approximately a hundred times greater than the observed fracture strength. Small cracks within a material can act as points of high stress-concentration and so reduce the nominal fracture strength. Griffith calculated that an internal elliptical crack of length about 10^{-6} m and with a tip radius of about 5×10^{-10} m would give a stress-concentration factor of about 100, and postulated that cracks of this order of size exist within glasses. Cracks of this order of size are too small to be seen, but if the Griffith crack theory is valid then a fine drawn glass fibre of diameter $< 10^{-6}$ m would be too small to contain such a crack normal to its axis, and hence should be extremely strong in tension. Fine drawn glass fibres are extremely strong and their good properties make them commercially useful (see section 8.12).

8.11 Polymers under Stress

The behaviour of polymeric materials, when subjected to stress, varies very considerably from one material to another. The structure of the polymer has a considerable effect on the properties of the material.

A linear polymer is composed of long fibrous molecules. Although termed linear, the individual molecules are not normally in straight-line form. The bond angle between adjacent covalently bonded carbon atoms is 109.5° and so the chain molecule may be randomly twisted or coiled. In addition, it is possible for rotation to occur at covalent bonds so giving some random molecular movement. This is the so called 'rubber' state. At much lower temperatures the same polymer would be in a rigid condition. The lower kinetic energy of the molecule means that bond rotation cannot occur to any great extent and adjacent molecules cannot easily move relative to one another. This is the 'glass' state. The modulus of elasticity, E, changes considerably at the transition from glass to rubbery state. The glass transition temperature, T_g, varies from one material to another and is lower for simple flexible chain molecules than for molecules with attached side groups. T_g for polyethylene is $- 120°C$ while T_g for polystyrene is $100°C$. The value of T_g for a polymeric material will be altered if a plasticiser is added to the material. Some polymeric materials are not completely amorphous and possess some crystallinity (see section 3.10) but these will still show a glass–rubber transition.

The term rubbery state includes the state in which a polymer may extend to many times its original length when stressed in tension, and return to its original length when the stress is removed, for example as the simple elastic band does. The term also includes the state in which the polymer strains under stress by a process of viscous flow, which produces considerable plastic deformation. Only a few polymers behave in the former manner.

If a polymer is stressed at a temperature well below its glass transition temperature, it behaves as a brittle solid obeying Hooke's law and fracturing with no plastic deformation. At temperatures closer to T_g some plastic yielding will occur before fracture. At a slightly higher temperature the shape of the stress–strain curve changes and the phenomenon of cold drawing occurs (see figure 8.11). At a stress corresponding to point Y on the curve the testpiece necks down considerably and thereafter considerable straining takes place at constant stress with undrawn material being drawn into the necked zone. The drawn material is much stronger than the undrawn material. During drawing the amorphous material becomes orientated with the chain molecules tending to lie along the direction of draw. The phenomenon of cold drawing can also occur in partially crystalline polymers, and with these materials it may occur at temperatures both above and below T_g.

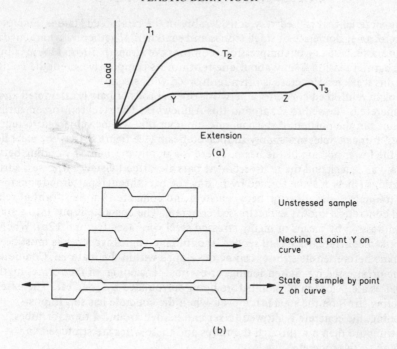

FIGURE 8.11   *Extension of thermoplastics: (a) typical load–extension curves for a thermoplastic, $T_1 < T_2 < T_3$; (b) cold drawing during tensile test at temperature T_3*

As can be seen the properties of linear polymers are extremely sensitive to temperature. Also they are sensitive to strain rate. An increase in the rate of straining results in an increase in the apparent strength of the material and a decrease in the amount of plastic deformation before fracture.

Not all polymeric materials are based on linear molecules. There are also network molecular compounds. The phenol-formaldehyde resins (bakelite) are an example of this type. Polymerisation causes the formation of a three-dimensional network molecular structure and this gives rise to a strong, but brittle, material that deforms under stress in a Hookean manner and ultimately fractures with no plastic deformation.

8.12 Reinforced Materials

There are very many situations in engineering where no single material will be suitable to meet a particular design requirement. However, two materials in combination may possess the desired properties, and provide a feasible solution to the materials-selection problem. The principle of composite materials is not new. The use of straw in the manufacture of dried mud bricks, and the use of hair and other fibres to strengthen plasters, date back to ancient civilisations.

One material that is used very considerably in the reinforced state is concrete. Concrete, an agglomerate of small stones and sand held together by a hardened cement paste, is strong in compression (compressive strength 50–65 MN/m^2) but its strength in tension is only about one-tenth of its compressive strength. To overcome this weakness concrete is often reinforced with steel.

In plain reinforced concrete, a network of steel rods or bars is assembled and the concrete is allowed to set around this framework. The steel reinforcement is positioned in the portion of the concrete member that will be subjected to tensile stresses. For example, in a simply supported beam (see figure 8.12a) the steel lies along the lower portion of the beam. There is a purely mechanical bonding between concrete and steel, and the reinforcement bars are often twisted, or possess surface projections (these may be formed by rolling the bars through patterned rolls) in order to increase the adhesion between steel and concrete. Another form of reinforced concrete is known as prestressed concrete. The concrete is put into a state of compression by means of highly stressed steel wires (see figure 8.12b). When a prestressed concrete beam is in service, the initial compressive stresses must be overcome before tensile stresses can be developed within the material. Concrete may be prestressed by pre-tensioning, or by post-tensioning. In the former method steel wires are placed in tension before being surrounded by concrete. The externally acting stress on the steel is removed when the concrete has set. In post-tensioning, the concrete is allowed to set and harden around a tube, or tubes. Steel wires are then put through the tubes and these wires are stretched and anchored to the concrete.

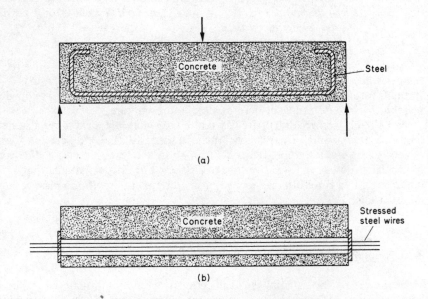

(a)

(b)

FIGURE 8.12   (a) *Plain reinforced-concrete beam − the upper layers of the beam are in compression and the lower layers are in tension; steel bars help to sustain the tensile stress;* (b) *prestressed concrete beam − steel wires in tension exert a compressive force on the concrete*

Glass is a material with a comparatively low strength, except when it is the form of very fine drawn fibres (see section 8.10). High-strength glass fibres are used in conjunction with polymers to form glass-reinforced plastics (fibreglass).

Very hard and strong materials are generally brittle, while the soft but very tough materials tend to have low yield strengths. In a fibre-reinforced material, high-strength fibres are encased within a tough matrix. The functions of the matrix are to bond the fibres together, to protect them from damage, and to transmit the load from one fibre to another. The greatest reinforcing effect is obtained when fibres are continuous and parallel to one another, and maximum strength is obtained when the composite is stressed in tension in a direction parallel to the line of the fibres.

Glass-fibre-reinforced plastics, have been in use as constructional materials for about 30 years. The glasses that are used are mainly s-glass (a soda–lime–silica glass) and e-glass (a calcium–alumino-borosilicate glass). The glass fibres are usually prepared with filament diameters of about $10\,\mu m$. The glass fibres are embedded in a matrix that is usually a polyester or an epoxide resin. Another reinforcing material that has been devloped in recent years is carbon fibre. Carbon fibre possesses a very high modulus of elasticity, and carbon fibres have been used successfully in conjunction with epoxide resins to produce low-density composites possessing high strength and modulus values. Carbon-fibre composites have been used for the manufacture of components for gas-turbine engines and rocket motors.

9

Phase Diagrams

9.1 Phases and Components

Matter can exist in three states, as solid, liquid or gas, and these three states are interdependent. In addition some solid substances can exist in more than one form, for example, there are two crystalline forms of carbon, namely graphite and diamond. Similarly, iron can exist in two crystalline forms, body-centred cubic and face-centred cubic.

The relationships between the various states of a substance, and the effects of temperature on these states can be shown by means of diagrams known as phase, or equilibrium diagrams. Phase diagrams may also be used to show the relationships between two or three substances, as for example between the component metals in any alloy system.

A *phase* may be defined as a portion of matter that is homogeneous. Mechanical sub-division of a phase will produce small portions indistinguishable from one another. A single phase is not necessarily a single substance. The gaseous state is always a single phase, irrespective of the number of gases present, since gases mix freely with one another in all proportions. Similarly, an unsaturated solution of salt in water is homogeneous and is therefore a single phase, but if the amount of salt is increased beyond the saturation limit the system will consist of two phases with a saturated solution of salt in water existing in equilibrium with excess solid salt.

Phase systems may be classified as one-component, two-component (binary) or three-component (ternary). The number of *components* in a phase system is the smallest number of atomic or molecular species needed to specify all the phases of the system. This statement needs to be clarified by examples. The phase system of ice, water and water vapour is a one-component system, the component being water, H_2O. The fact that water is a compound of hydrogen and oxygen does not affect matters since water does not dissociate into its constituents under normal conditions. In the case of the alloy system of the metals copper and zinc there are six different solid phases formed, each possessing a different crystal structure, but this is a two-component system since all the phases can be expressed in terms of copper and zinc.

9.2 The H₂O Diagram

A system consisting of a pure substance, a one-component system, may be represented by a phase or equilibrium diagram with pressure and temperature as the two axes (it is customary to plot pressure as the ordinate). The phase diagram for the solid, liquid and vapour phases of water is shown in figure 9.1.

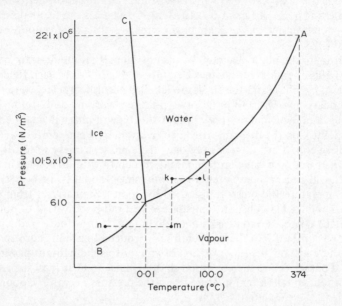

FIGURE 9.1 *Phase-diagram for water*

The curve OA represents the variation of vapour pressure of water with temperature. Similarly the curve BO indicates the variation of vapour pressure of ice with temperature. The curve OC represents the pressures and temperatures at which water and ice are in equilibrium, or in other words it indicates the effect of pressure on the melting point of ice. For clarity the slope of the line OC is exaggerated. If the pressure on a solid and liquid in equilibrium is increased the phase with the larger volume will tend to disappear. This is in accordance with Le Chatelier's principle, which may be stated as follows. *If, for a system in equilibrium, one of the factors such as temperature or pressure is changed then the position of equilibrium will shift in an attempt to offset the effect of the change.* Ice is less dense than water and so an increase in pressure will lead to a decrease in melting point. An increase of one atmosphere (10^5 N/m²) in pressure will reduce the freezing point of water by 0.0075°C. (With most substances the liquid phase is less dense than the solid and an increase in pressure would cause an increase in melting point.)

The curve OA shows that an increase in temperature causes the vapour pressure of water to rise. A liquid is said to boil when its vapour pressure is equal to the external pressure. Point P on the curve OA represents the normal boiling point of water at 100.0°C (373.13 K) and a pressure of 101.5×10^3 N/m² (1 atmosphere).

Ice also has a vapour pressure and this, although small, is shown by curve BO. BO is not a continuation of the curve OA, but is a separate curve since it refers to a separate phase. The two vapour pressure curves intersect at O. Point O is a *triple point* since three phases, water, ice and vapour, exist in equilibrium. The vapour pressure of the two phases at the triple point is $610 \, N/m^2$. The normal melting point of a solid and the triple point are not coincident. The normal melting point of a solid is the temperature at which the solid melts at atmospheric pressure. For ice the melting point occurs at $0°C$ (273.13 K), whereas the triple point, where ice is in equilibrium with water under the pressure of its own vapour, occurs at $+ 0.01°C$ (273.14 K).

Referring to the phase diagram for water (figure 9.1), the diagram consists of areas or fields bounded by the lines OA, OB, and OC. Within each field there is only one stable phase and the fields are labelled accordingly, ice, water and vapour. At a boundary line, OA, OB or OC, two phases coexist in equilibrium.

As an example of the interpretation of this type of phase diagram consider the state at point k on the diagram with specified values of temperature and pressure. Under these conditions there is only one stable phase, namely water. If the temperature is increased to a value corresponding to point l with no change in pressure the liquid will convert completely to vapour, since vapour is the only stable state at the temperature and pressure specified by point l. Similarly, if from state k the pressure is reduced at constant temperature to a value equivalent to point m there will again be complete vapourisation of the liquid. If from state m the pressure is maintained constant, but the temperature is reduced to a value corresponding to point n there will be direct conversion of vapour to solid without passing through the liquid phase. This is the condition that gives rise to hoar-frost deposition, namely a sudden fall in temperature when the pressure of water vapour in the atmosphere is less than $610 \, N/m^2$.

Point A in the diagram is the *critical point*. Beyond this point liquid and vapour phases become identical. The critical point is invariant and for water has specific values of temperature and pressure ($374°C$ (647 K) and $22.1 \times 10^6 \, N/m^2$).

The change from vapour to liquid, or from liquid to vapour generally occurs quite suddenly. A series of isothermal curves is shown in figure 9.2. At some low temperature T_1 the behaviour of a gas departs considerably from 'ideal' behaviour as stated by the relationship

$$pV = RT$$

As pressure is increased at temperature T_1 the volume of gas decreases following the curve AB but at a pressure corresponding to point B the volume suddenly reduces to the low value C as the gas liquifies. A further increase in pressure causes little further reduction in volume since the liquid is not very compressible. This is shown by the portion CD on the T_1 isothermal. In following the path ABCD at a constant temperature the transition from gaseous to liquid state occurs suddenly. It is possible to proceed from A to D by a different route. If the temperature of the vapour is increased from T_1 to a high temperature, keeping the volume constant, the pressure will rise appreciably, following the path AP. If the pressure is now maintained at a constant value and the temperature reduced to T_1 the path PD will be followed and the substance will have transformed from vapour to liquid, but in a gradual manner with no sharp discontinuity. This indicates that under these

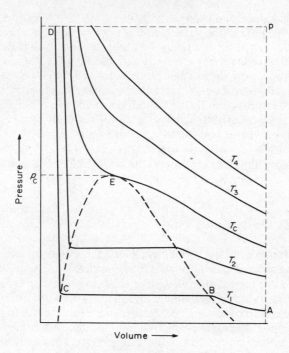

FIGURE 9.2 *Isothermal curves for a substance;* $T_1 < T_2 < T_c < T_3 < T_4$

conditions there is no difference between gas and liquid. This is referred to as *the continuity of the liquid and gaseous states.* At the critical point E on the critical-temperature isothermal the densities of liquid and saturated vapour are identical.

All gases show this type of behaviour but values of critical pressure and critical temperature vary considerably from one substance to another.

The phase diagrams for other pure substances are similar to that for water. The phase diagram for carbon dioxide shows that the triple point pressure is 520×10^3 N/m^3 (approximately 5 atmospheres). The vapour pressure of solid carbon dioxide, 'dry ice', is equal to standard atmospheric pressure at a temperature of $-78°C$. At atmospheric pressure solid carbon dioxide *sublimes*, that is, it transforms directly from the solid state into the vapour phase without liquefying. Any substance will sublime rather than melt when heated at atmospheric pressure if the triple point pressure is higher than atmospheric.

9.3 Polymorphism

Certain solid substances can exist in more than one crystalline form. This is termed *polymorphism* or *allotropy*. Among the elements that exhibit allotropy are carbon, iron, sulphur and tin. Diamond and graphite are two allotropic forms of carbon.

Of the number of metals that show allotropic modifications, the most commercially important is iron. Iron can exist in two crystalline forms, body-centred cubic and face-centred cubic. The crystal structure of iron is body-centred cubic at all temperatures from zero up to 908°C (1181 K). This form is termed α-iron. On heating beyond 908°C the structure of iron changes to face-centred cubic, a more closely packed and hence denser state. This form is termed γ-iron. γ-iron remains the stable form up to 1388°C (1661 K) when the structure reverts to the body-centred cubic form. The high temperature body-centred cubic structure is termed δ-iron, but it is crystallographically identical with α. The δ-iron is stable at temperatures up to the melting point of 1535°C (1808 K). α-iron loses its ferromagnetic characteristics on heating above 768°C (1041 K) and early workers used the term β to describe the state of iron at temperatures between 768°C and 908°C. When it was discovered that there was no crystallographic change associated with the loss of magnetism use of the term β-iron was discontinued.

9.4 Metallic Alloy Systems

An alloy is a mixture of two or more metals, or a mixture of a metal and a non-metal, with the mixture exhibiting metallic properties. Most alloys are made in the liquid phase and it is convenient to consider the formation of alloy structures on the basis of the solidification of liquid alloys.

When two liquid metals are mixed together they may dissolve in one another in all proportions, forming a homogeneous liquid solution, or they may be insoluble in one another, either partially or totally, and separate out into two liquid layers, like oil and water. Since virtually all commercial alloy systems are based on systems in which the liquid metals are completely soluble in one another we shall not consider the other possibilities in this book.

Although two metals may be completely soluble in one another in the liquid phase this does not necessarily mean that they will solidify to give a homogeneous solid phase, or solid solution. The two metals may be

(1) totally insoluble in one another when solid
(2) totally soluble in one another when solid
(3) partially soluble in one another when solid, or
(4) combine with one another to form an intermetallic compound.

The type of alloy structure, and hence the alloy properties obtained, will depend on the nature of the solid phase or phases formed. A convenient way of presenting this information is to use a phase, or equilibrium diagram. For a binary system involving the two components, A and B, the equilibrium diagram is conventionally drawn with a base line representing composition, and the ordinate representing temperature. The base line will range from 100 per cent A, 0 per cent B at the left to 0 per cent A, 100 per cent B on the right-hand side, and compositions may be expressed either as weight percentages or as atomic percentages. Of these, the former is the more usual system.

If it is desired to represent a system containing three metals A, B, and C, the base of the equilibrium diagram must be a triangular plane. Temperature is then represented on an axis perpendicular to this plane and the completed ternary equilibrium diagram is a three-dimensional solid figure. Only binary phase diagrams will be discussed in this book.

9.5 Binary Alloy Phase Diagrams

Consider first the case of two pure metals that are totally soluble in one another when liquid but totally insoluble in one another in the solid state.

In the same manner as the presence of dissolved salt will depress the freezing point of water, so the freezing point of a liquid metal will normally be depressed if the liquid metal contains some other substance in solution. The phase, or equilibrium, diagram for such a binary system of two pure metals, A and B, is shown in figure 9.3. Point A in this diagram denotes the freezing temperature of pure metal A and the depression of freezing point of A containing dissolved metal B is shown by the curved line AE. Similarly, the freezing point of pure metal B is shown by point B on the diagram and the curve BE indicates the effect of dissolved A on the freezing point of pure B. The two freezing-point curves intersect at point E. Point E is termed the *eutectic* point and it represents the lowest temperature at which a liquid solution of A and B can exist.

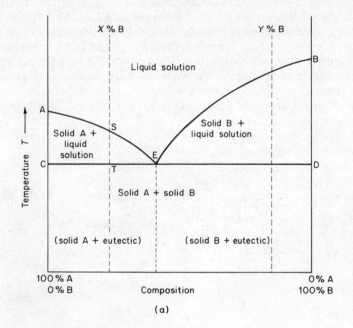

FIGURE 9.3 *Binary phase diagram for solid insolubility (simple eutectic)*

Consider the freezing of a liquid alloy containing X per cent of metal B. The presence of X per cent of metal B dissolved in A will depress the freezing point of metal A to a value given by point S on the curve AE. Freezing will commence at this temperature with the separation of crystals of solid pure metal A. If pure A is rejected from the solution the composition of the remaining liquid must become enriched in B, that is, the composition of the liquid varies toward the right. This means that as the freezing of A continues the temperature and composition of the liquid remaining follows the curve AE toward point E. Point E, which is the only point common to both freezing point curves, represents the lowest temperature that a liquid solution can exist at, and at this point all remaining liquid solution solidifies forming a fine-crystal-grained mixture of both solids A and B.

The fine-grained crystal mixture formed is termed the eutectic mixture. The final structure of the solid mixture containing X per cent of B will therefore be composed of large crystals of pure A (primary crystals) and a eutectic mixture of A and B.

If a liquid solution containing Y per cent of B is allowed to solidify, solidification would follow a similar pattern, but in this case primary crystals of pure B would solidify first. It is important to note that the composition of the eutectic mixture remains constant.

In the phase diagram line AEB is called the *liquidus* and line CED is called the *solidus*. At all points above the liquidus the mixture is always liquid, and below the solidus the mixture is always wholly solid. Between liquidus and solidus, in the solidification range, the mixture is in a pasty stage.

Many metals solidify from liquid in a dendritic manner. Solidification commences at a nucleus and outward growth from the nucleus occurs preferentially in three directions. Subsequently secondary and then tertiary arms grow producing a skeleton-type crystal, as in figure 9.4. Outward growth ceases when the advancing dendrite arms meet an adjacent crystal. When outward growth has ceased the dendrite arms thicken and eventually the whole mass is solid and no trace of the dendritic formation remains, except where shrinkage causes interdendritic porosity, or in alloy systems where the final liquid to solidify is of a different composition from the primary dendrites.

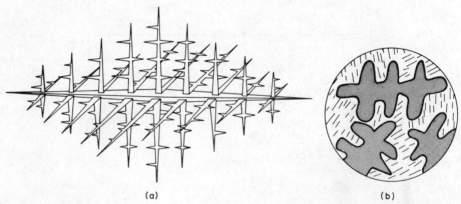

(a)   (b)

FIGURE 9.4   (a) *Representation of a dendrite;* (b) *solid structure of a simple eutectic alloy — dendrites of A in eutectic mixture*

It is possible for metals to form what is termed a *solid solution*. This concept may seem strange, but it simply means that the atoms of the two elements have taken up positions in a common crystal lattice forming a single phase. The atoms of one element enter into the space lattice of the other element in either an *interstitial* or *substitutional* manner, as in figure 9.5 (see also section 8.3). The arrangement of dissolved atoms is normally random, but in some instances substitutional solid solutions of an ordered type may be formed. An ordered solution (also known as superlattice) can only exist at one fixed composition. If the two metals are of the same crystal type and their atomic diameters are within 7 per cent of one another, it is possible for there to be complete solid solubility over the whole range of composition. When these conditions are not satisfied then solid solubility, if it occurs at all, will be restricted.

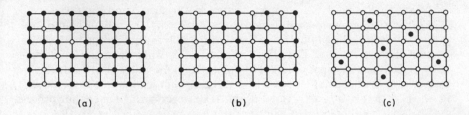

(a)　　　　　　　　(b)　　　　　　　　(c)

FIGURE 9.5 *Schematic representation of solid solution:* (a) *substitutional (random);* (b) *substitutional (ordered);* (c) *interstitial*

For a binary alloy system where there is a continuous range of solid solution formed the possible phase-diagram shapes are as shown in figure 9.6. It should be noted that point C in figure 9.6b is not a eutectic point.

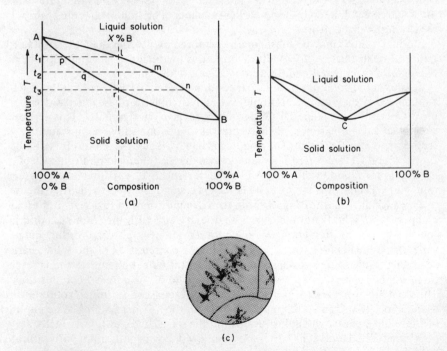

FIGURE 9.6 (a) *Phase diagram for complete solid solubility;* (b) *alternative phase diagram for this alloy type;* (c) *cored crystal structure*

An alloy containing X per cent of B (see figure 9.6a) would solidify in the following manner. Freezing of the liquid solution would commence at temperature t_1. At this temperature liquid of composition l would be in equilibrium with a solid solution of a composition corresponding to point p on the solidus, so the first solid solution crystals to form are of composition p. Consequently the composition of the remaining liquid becomes enriched in the metal B and the freezing temperature falls slightly. As the temperature falls so the composition of the solid solution tends to change by a diffusion process following the solidus line toward B. At some temperature t_2 liquid of composition m is in equilibrium with solid solution of composition q. Solidification of this alloy will be complete at temperature t_3 when the last drops of liquid of composition n solidify, correcting the composition of solid-solution crystals to r. If the solidification rate is very slow, allowing for the attainment of equilibrium at all stages during the cooling process, the final solid-solution crystals will be uniform in composition. In practice, however, solidification rates are too rapid for full equilibrium to be attained and the crystals will be *cored*. In a cored crystal the composition is not the same at all points. The crystal lattice is continuous but there will be a gradual change in composition across each crystal. The centre of a crystal will be rich in metal A while the outer edges will be rich in metal B. In some alloy systems the coring in crystals is clearly visible under microscopical examination. With alloys of copper and nickel, for example, where the alloy colour is dependent on composition, the centres of crystals are rich in nickel and silvery in appearance and the outer edges of crystals are rich in copper and darker in colour. This colour shading clearly shows the dendritic manner of growth.

Coring in alloys may be subsequently removed by heating the material to a temperature just below the solidus. During this treatment (annealing), diffusion takes place, evening out composition gradients within the crystals.

It is far more common to find that solid metals are partially soluble in one another rather than totally insoluble or totally soluble. A phase diagram for a binary system showing partial solid-solubility is given in figure 9.7. This diagram is in effect a combination of the two previous types and shows solid-solubility sections and also a eutectic. The liquidus is line AEB and the solidus is ACEDB. Lines FC and GD are *solvus* lines and denote the maximum solubility limits of metal B in metal A and of metal A in metal B respectively. As there are two separate solid solutions formed, the Greek letters α and β are used to identify them.

Consider the solidification of three alloy compositions in this system. For alloy composition 1 solidification begins at temperature t_1 with the formation of β solid solution of composition O. As cooling continues the composition of the liquid varies along the liquidus toward point E and the composition of the solid β varies according to the solidus toward point D. When the eutectic temperature is reached there will be primary cored crystals of β and liquid of the eutectic composition. This liquid then freezes to form a eutectic mixture of two saturated solid solutions, α of composition C and β of composition D. During further cooling the compositions of the α and β phases will adjust, following the solvus lines, until eventually at point p saturated α crystals of composition q will be in equilibrium with saturated β solid solution of composition r.

For alloy 2 solidification of the liquid solution takes place in the same manner as for a complete solid-solution alloy and when solidification is complete the structure will be one of cored α crystals.

In the case of alloy 3 a new concept emerges, namely the possibility of structural changes occurring within the solid state. The liquid alloy will freeze on cooling to give a cored α solid solution. During further cooling below the solidus the α solid solution will remain unchanged until temperature t_2 is reached. At this temperature the composition line meets the solvus and the solid solution is fully saturated with metal B. As the temperature of the alloy falls below t_2 the solubility limit is exceeded and excess metal B is rejected from solution in A as a precipitate. In this case it is not pure metal B which forms as a second solid phase, but rather, saturated β solid solution. Eventually at temperature t_3 the structure is composed of α crystals of composition q with precipitated particles of composition r. The second phase, β, may be precipitated either at the α grain boundaries, within the α crystals, or at both types of site (see figure 9.7b).

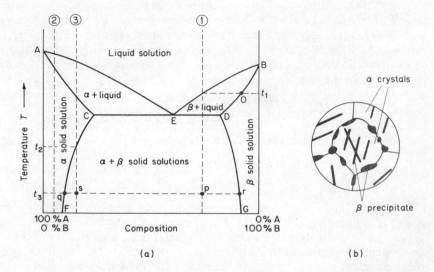

(a)   (b)

FIGURE 9.7   (a) *Phase diagram for partial solid solubility with eutectic;* (b) *structure of alloy 3*

Changes within the solid state take place slowly compared with changes between liquid and solid states. In consequence they may be suppressed by rapid cooling. Rapid cooling of alloy 3 from some temperature below the solidus may prevent the precipitation of β from taking place and giving at temperature t_3 an α solid solution of composition s supersaturated with dissolved metal B. This is of significance in connection with precipitation hardening and age hardening and will be discussed further in section 9.6.

It is appropriate at this stage to bring in some simple rules for the interpretation of phase diagrams.

(1) A phase diagram consists of lines that divide it into a number of areas, or fields. These fields may be single-phase, as is the area bove the liquidus in figure 9.3, or two-phase, as in area ACE in the same figure.
(2) Single-phase areas are always separated by a two-phase zone, and three phases can only coexist at a point, such as the eutectic point.

(3) When a vertical line representing the composition of some alloy in the system crosses a line in the phase diagram, it indicates that some change is taking place in the alloy. For example, in figure 9.3 the X per cent B line cuts two other lines, AE at S and CD at T, indicating that during cooling the alloy is beginning to freeze at point S and that solidification is complete at point T.

(4) For any point in a two-phase region the composition of the two phases in equilibrium with one another can be determined. If a horizontal line is drawn through the point the intersections of this line with the phase-boundary line denotes phase compositions. For example, in figure 9.6 for an alloy containing X per cent B at temperature t_2 the intersections at q and m indicate that a solid solution of composition q is in equilibrium with a liquid solution containing m per cent of metal B.

(5) The relative proportions of the phases present can be determined using the lever rule. The quantitites of phases present are in proportion to the lengths of the lever lines, for example in figure 9.7 for alloy composition 1 at temperature t_3 the phases present are an α solid solution of composition q and a β solid solution of composition r in the ratio

$$\frac{\text{quantity of } \alpha \text{ solid solution}}{\text{quantity of } \beta \text{ solid solution}} = \frac{\text{pr}}{\text{qp}}$$

Similarly for alloy composition 3 at temperature t_3

$$\frac{\text{quantity of } \alpha \text{ solid solution}}{\text{quantity of } \beta \text{ solid solution}} = \frac{\text{sr}}{\text{qs}}$$

The physical and other properties of solid alloys are greatly influenced by the type of alloy phase-diagram involved. For systems in which the component metals are completely insoluble in one another in the solid state, the structure of the solid alloy is simply a mixture of two pure metals. Consequently, the variation of properties with alloy composition should be linear. In actual practice there is a departure from linearity due to a grain-size effect. A eutectic is a finely divided mixture of two metals. The primary crystals that solidify first on either side of the eutectic point are much larger in size. A fine-crystal-grained metal tends to be harder and stronger than a coarse-grained sample of the same material. Similarly a fine grain-size causes a reduction in electrical and thermal conductivities. In figure 9.8a the approximate relationship between two properties, hardness (H) and electrical conductivity (G), and alloy composition is shown for a simple eutectic alloy. The dotted lines show the expected property variation, neglecting the grain-size effect. Yield strengths and tensile strengths follow a similar pattern to hardness.

In a solid-solution alloy the presence of the solute atoms imposes strain in the parent lattice strengthening the alloy (see section 8.6). Maximum strengthening occurs when the lattice is subjected to maximum strain, that is, when there are equal numbers of both types of atoms. 50 per cent atomic is not necessarily the same as 50 per cent by weight. Property variations with composition for solid-solution alloys are shown in figure 9.8b.

Figure 9.8c shows the relationship between properties and composition for the partial solid solubility case. Since this phase diagram is a combination-type diagram, so the property diagrams are combinations of the former two types.

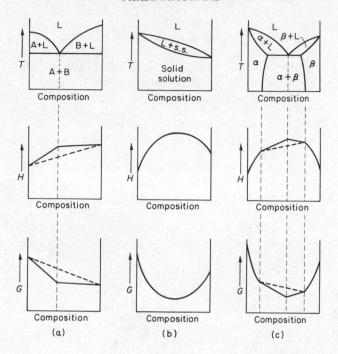

FIGURE 9.8   *Relationships between alloy composition and hardness (H) and electrical conductivity (G) for* (a) *simple eutectic,* (b) *solid solubility and* (c) *partial solid solubility with eutectic*

9.6 Precipitation Hardening

Certain metallic alloys can be strengthened by the process termed *precipitation hardening*. The phenomenon was first observed in aluminium–copper alloys by the German chemist Wilm at the beginning of this century. Wilm's observations led to the development of the high-strength heat-treatable aluminium alloys, frequently referred to as duralumin-type alloys.

Precipitation hardening is only possible if, by means of suitable heat treatments, an alloy can be placed in a metastable condition. The alloy systems in which this process may be possible are those in which there is partial solid solubility. Consider the aluminium-rich end of the aluminium–copper phase diagram (see figure 9.9a). Within this system the alloy compositions that respond best to precipitation hardening are those containing between 4 per cent and 5.5 per cent of copper. It will be noticed in figure 9.9a that the maximum solubility of copper in aluminium is 5.7 per cent (by weight) at the eutectic temperature 548°C but is only 0.2 per cent (by weight) at low temperatures. When an alloy containing, say, 5 per cent of copper is heated to 550°C all the copper present will be held in solid solution in the aluminium lattice. If the alloy is allowed to cool slowly from this temperature, equilibrium conditions will be established as the solvus line is crosssed and the second phase will be precipitated from the saturated solid solution. The second phase in this case is a compound, $CuAl_2$, known as θ phase. After slow cooling to room temperature the equilibrium structure will consist of a coarse precipitate of θ phase in a dilute solid solution of copper in aluminium. If, however, the alloy were to be rapidly cooled, by quenching in water, from 550°C, the whole of the copper content would be retained in solid solution within the aluminium. Such a

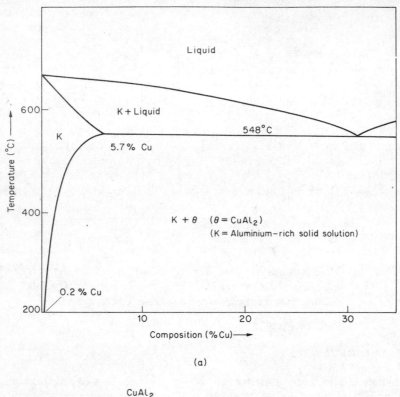

(a)

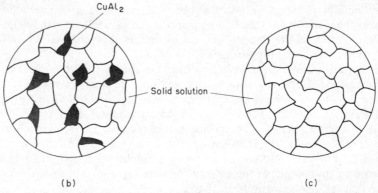

(b)   (c)

FIGURE 9.9   (a) *Aluminium-rich end of aluminium–copper phase diagram;* (b) *5 per cent* Cu *alloy slowly cooled from* 550°C *showing* $CuAl_2$ *precipitate;* (c) *5 per cent* Cu *alloy rapidly cooled from* 550°C *supersaturated solid solution*

solution, being highly supersaturated with dissolved copper, is a non-equilibrium phase and hence is metastable. It will possess a tendency to change into the stable structure of dilute solution plus θ precipitate. Before there can by any precipitation of θ phase there must be some diffusion of copper through the aluminium lattice in order to increase the copper concentration at some points. As this pre-precipitation diffusion takes place there will be a considerable increase in the amount of lattice strain at localised points within the aluminium lattice. This build-up of strain within the lattice would have the effect of causing an increase in the hardness and strength of the alloy.

The diffusion of one solid metal through another, and the process of precipitation are thermally activated processes and as such conform to the Arrhenius equation (section 2.3). The rate of diffusion of copper in aluminium, though slow, is sufficient for hardening to occur at 25°C, but cooling the alloy to 0°C is sufficient to halt the process. In a number of commercial aluminium–copper alloys containing small percentages of other elements the presence of the other elements reduces the copper diffusion rate to such an extent that it will not occur at 25°C (room temperature). Heating the alloys to some temperature above room temperature will increase diffusion rates and hence the rate at which the alloy hardens.

The hardness of the material continues to increase as the diffusion of copper proceeds. When the concentration of copper in the copper-rich areas has built up to the required level the compound $CuAl_2$ may be precipitated from solid solution. The formation of separate particles of $CuAl_2$ releases some of the strain within the aluminium crystal lattice and this causes a softening of the alloy. The true precipitation stage of $CuAl_2$ from supersaturated solution can only take place at elevated temperatures. There is insufficient energy available for this to occur at ordinary temperatures.

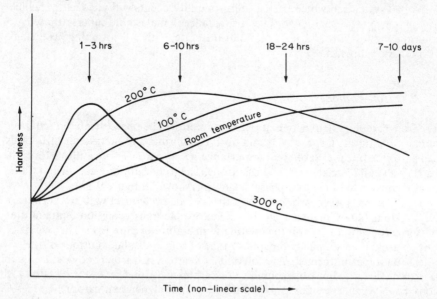

FIGURE 9.10 *Hardness–time relationships for an aluminium–copper age-hardening alloy*

Figure 9.10 shows the relationships between hardness, time and temperature for an alloy of aluminium and copper. When an alloy starts to harden spontaneously after receiving a solution heat treatment, that is, rapidly quenched from high temperature to give a metastable superstaturated solution, the process is termed age hardening. If, after solution heat treatment, it is necessary to heat an alloy to some temperature for diffusion and hardening to occur, this second heat treatment is termed precipitation heat treatment and the alloy is referred to as a precipitation-hardening alloy, even though the treatment is ceased before true precipitation and softening can occur. The softening at high temperature is termed *over-ageing*.

Any binary system for which the phase diagram shows partial solid solubility may possess alloy compositions that will respond to a precipitation-hardening process. There are other alloys based on aluminium that may be strengthened in this manner. Nickel is partially soluble in aluminium and alloys containing both nickel and copper are used extensively. The presence of nickel reduces diffusion rates within aluminium and in consequence Al–Cu–Ni alloys can be heated to higher temperatures than straight Al–Cu alloys before over-ageing occurs. These alloys are used for applications in which retention of strength at elevated temperatures is a requirement. They are used in the construction of the Concorde super-sonic aircraft where frictional heating occurs at high airspeeds and for the cylinder blocks and heads of internal-combustion engines. Magnesium and silicon form a compound, Mg_2Si, when added in the correct proportions and the phase diagram between this compound and aluminium shows partial solid solubility. Al–Mg–Si alloys will respond to precipitation hardening.

Other systems in which this process is used to strengthen commercial alloys include Cu–Be, Cu–Cr, several nickel alloys and certain alloy steels rich in nickel (maraging steels).

Not all systems in which partial solid solubility occurs will yield alloys which can be precipitation hardened. In some instances a metastable supersaturated solution will revert to the stable equilibrium state, with full softening occurring at ordinary temperatures.

9.7 The Iron–Carbon System

The iron–carbon system is one of extreme interest. The phase diagram contains both a eutectic and a *eutectoid*† and both equilibrium and metastable structures can be formed. It is also of great importance because iron–carbon alloys form the basis of all commercial steels and cast irons. The iron–carbon phase diagram, or more correctly the Fe–Fe_3C phase diagram, is shown in figure 9.11.

Iron combines with carbon to form the carbide *cementite*, with the formula Fe_3C. Along the ordinate are plotted the allotropic transformation temperatures of iron (see section 9.3) and the melting temperature of pure iron. The size of a carbon atom is very small compared to that of iron and solid solutions of the interstitial type are formed. The solubility of carbon in α-iron (body-centred cubic) is extremely limited but the solubility of carbon in γ-iron (face-centred cubic) is considerably greater reaching a maximum limit of 1.7 per cent carbon at 1130°C.

---

†A eutectoid is wholly within the solid state. A single-phase solid solution, when cooled through the eutectoid temperature, transforms into two other solid phases.

Names have been assigned to the various phases within the Fe–Fe$_3$C system, as
follows: the body-centred cubic phases, α and δ, are termed *ferrite*, the face-centred
cubic phase, γ, is termed *austenite* and the eutectoid mixture of α and cementite
is known as *pearlite*.

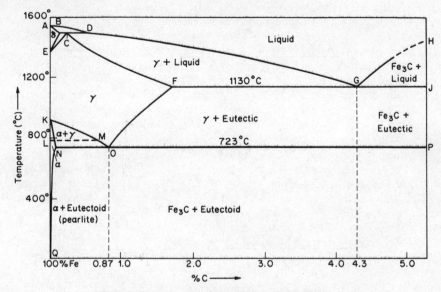

FIGURE 9.11   Fe–Fe$_3$C *phase diagram*

The full Fe–Fe$_3$C phase diagram may appear at first sight to be extremely complex but it can be divided into sections which, in themselves, are straightforward.
Steels are basically alloys of iron and carbon containing up to 1.5 per cent of
carbon. Therefore for the consideration of steels, and in particular their heat
treatment, it is convenient to consider only that portion of the diagram up to a
carbon content of 1.5 per cent and up to a temperature of 1000°C (see figure 9.12).

It will be seen that ferrite cannot hold carbon in solid solution to any great
extent, the limits being 0.04 per cent of carbon at 723°C and 0.006 per cent of
carbon at 200°C. Austenite, however, can hold a considerable amount of carbon
in solid solution, ranging from 0.87 per cent at 723°C to 1.7 per cent at 1130°C.
The eutectoid point occurs at a temperature of 723°C and at a carbon content of
0.87 per cent. The terms hypoeutectoid and hypereutectoid are used to denote
steels which contain less carbon than, and more carbon than, the eutectoid
composition, respectively.

The presence of carbon depresses the α–γ transformation temperature of iron.
Line KMO in figures 9.11 and 9.12 denotes this transformation temperature, and
its dependence upon composition. Lines OF and QN are solvus lines and denote
the maximum solubility limits of carbon in γ- and α-iron respectively. Point O is
the eutectoid, or pearlite, point. The line LMOP indicates the Curie temperature,
at which loss of magnetism occurs on heating.

If a sample of a steel is heated or cooled, and accurate measurements are taken,
thermal arrest points will be noted, corresponding to the phase transformation
lines (and Curie temperature) on the phase diagram. The phase line NOP is known

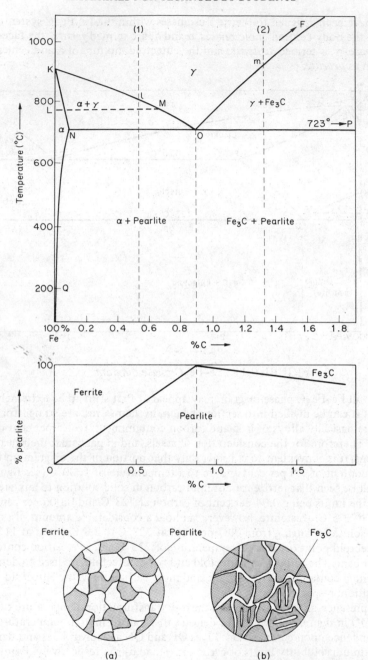

FIGURE 9.12 *Steel portion of the* Fe-Fe$_3$C *diagram:* (a) *microstructure of hypoeutectoid steel (1);* (b) *microstructure of hypereutectoid steel (2)*

as the A_1 transformation, the Curie temperature, LM as the A_2 transformation, the phase line KMO as the A_3 transformation, and the line OF as the A_{cm} transformation. These arrest points, or transformation temperatures, are also known as the *critical points*, or *critical temperatures*, for the steel. The eutectoid temperature, A_1, is known as the lower critical temperature, and the α–γ transformation, line KMO, is known as the upper critical temperature. If a steel is heated or cooled very slowly, so that equilibrium conditions are approached, the measured arrest temperatures will agree with the values shown on the iron–carbon phase diagram. With more rapid heating or cooling rates, the measured arrest points will differ from equilibrium values. They will be higher than the equilibrium values when determined during heating, and lower when determined during cooling. Values measured during heating are written as A_{c_1}, A_{c_2}, and A_{c_3}, while values determined during cooling are written as A_{r_1}, A_{r_2}, and A_{r_3}.†

Let us now consider the changes that occur during the cooling of steels of various compositions. Refer to figure 9.12, and consider first the cooling of a hypoeutectoid steel of composition 1. At a high temperature the steel structure will be composed of homogeneous crystals of austenite solid solution. On cooling to the upper critical temperature, point 1 on the diagram, austenite will begin to transform into ferrite. The ferrite can hold very little carbon in solid solution and so the remaining austenite becomes enriched in dissolved carbon. As the temperature falls, more ferrite is formed, and the composition of the remaining austenite increases in carbon content, following the line KMO. When the lower critical temperature is reached, the austenite, which is now of eutectoid composition, transforms into the eutectoid mixture pearlite, a mixture composed of alternate layers of ferrite and cementite.

For a hypereutectoid steel of composition 2 the homogeneous austenite structure that exists at high temperatures will begin to change when the temperature has fallen to a point m on the solvus line OF. This is the saturation limit for dissolved carbon in austenite and on cooling below the temperature of point m, excess carbon precipitates from solid solution in the form of cementite. The cementite appears in the microstructure as a network around the austenite crystals, and also in the form of needles within the austenite crystal grains. The carbon content of the austenite reduces with further cooling and when the lower critical temperature is reached all remaining austenite, which is now of eutectoid composition, transforms into pearlite. The presence of cementite in the form of needles, or as a boundary network, renders the steel brittle, and heat treatment is necessary to put the steel into a suitable condition for many applications.

Ferrite is a comparatively soft and ductile constituent possessing a tensile strength of about 280 MN/m^2. The tensile strength of pearlite formed by slow cooling from the austenitic range is about 700 MN/m^2, but its ductility is very much less than that of ferrite. It must be emphasised that the properties quoted only apply to slowly cooled steels. An increase in the rate of cooling through the critical temperature range will alter the structure, and hence the properties, of any steel. When the cooling rate is increased, there is some undercooling of austenite to below the equilibrium transformation temperatures. Once the phase change from undercooled

---

†The symbol A signifies 'arrest' (French arriere). The subscripts c and r are from the French: chauffage – heating; refroidissement – cooling.

austenite to pearlite commences it takes place very rapidly resulting in very fine lamellae of ferrite and cementite. The hardness and strength of pearlite is dependent on the interlamellar spacing, and very fine pearlite formed by rapid cooling may have tensile strengths of the order of 1300 MN/m².

If a steel is cooled extremely rapidly there will be insufficient time allowed for austenite to decompose into pearlite and, instead, the austenite changes into a metastable body-centred lattice with all the carbon trapped in interstitial solid solution. Ferrite can theoretically hold virtually no carbon in solution. The rapidly cooled structure that is formed is highly strained and distorted by the large amount of dissolved carbon into a body-centred tetragonal lattice. This constituent is termed *martensite*, and it is extremely hard and brittle. The hardness of martensite depends on the carbon content, and is greatest in high carbon steels, that is, the greatest degree of lattice strain. Martensite under a microscope appears as a series of fine needle-like (acicular) crystals. A martensitic structure can be formed by rapidly quenching a heated steel from the austenitic state in water or oil. This is the treatment termed hardening. Martensite is a non-equilibrium phase that does not appear in the iron–carbon phase diagram.

Martensite, although very hard, is also extremely brittle and a hardened steel requires a further heat treatment, known as tempering, before it can be put into service. When the metastable martensitic structure is heated it becomes possible for the carbon trapped in supersaturated solid solution to diffuse through the lattice and precipitate from solution in the form of iron-carbide particles. This precipitation will relieve the strain within the lattice and cause the hardness and brittleness of the material to be reduced. This diffusion process can commence at temperatures of about 200°C, but the rate of diffusion is extremely slow at this temperature. An increase in the temperature will cause an increase in diffusion and precipitation rates and therefore increase the extent of the softening. At temperatures up to 450°C the carbide precipitate particles are much too fine to be resolved under the optical microscope, although their presence may be detected by using more sophisticated techniques. At higher temperatures the carbide particles increase in size, and at 700°C the cementite coalesces into a series of fairly large, and roughly spheroidal particles (700°C is just below the lower critical temperature). This gives rise to a soft, but extremely tough, material. When microscopy was first used to investigate the changes that take place during the tempering of quenched steels, the terms *troostite* and *sorbite* were assigned to the distinctive types of structures produced by tempering at temperatures in the region of 400°C and 500°C respectively. These terms still remain in use although the structures could be more properly described as tempered martensite. Tempering temperatures and times have to be fairly accurately controlled in order to produce the desired properties in the material.

As stated earlier the hardness and strength of steels vary very considerably with both carbon content and type of heat treatment. Certain names that relate to the carbon content are used in connection with steels. *Mild steels* are those containing up to 0.3 per cent of carbon. Steels containing between 0.3 and 0.6 per cent of carbon are termed *medium-carbon steels*, and these may be hardened and tempered. Steels containing more than 0.6 per cent of carbon are always used in the hardened and tempered condition, and these are known as *high-carbon steels*, or *tool steels*. Table 9.1 gives some typical carbon contents and uses of steels.

Properties of steels may be modified and improved by the addition of other alloying elements, including chromium, manganese, nickel, silicon and tungsten.

TABLE 9.1

Compositions and typical applications of steels

C (%)	Name	Applications
0.05	Dead mild steel	Sheet and strip for presswork, car bodies, tin-plate; wire, rod, and tubing
0.08–0.15	Mild steel	Sheet and strip for presswork; wire and rod for nails, screws, concrete-reinforcement bar
0.15	Mild steel	Case-carburising quality
0.1–0.3	Mild steel	Steel plate and sections, for structural work
0.25–0.4	Medium-carbon steel	Bright drawn bar
0.3–0.45	Medium-carbon steel	Shafts, and high-tensile tubing
0.4–0.5	Medium-carbon steel	Shafts, gears, railway tyres
0.55–0.65	High-carbon steel	Forging dies, railway rails, springs
0.65–0.75	High-carbon steel	Hammers, saws, cylinder linings
0.75–0.85	High-carbon steel	Cold chisels, forging die blocks
0.85–0.95	High-carbon steel	Punches, shear blades, high-tensile wire
0.95–1.1	High-carbon steel	Knives, axes, picks, screwing dies and taps, milling cutters
1.1–1.4	High-carbon steel	Ball-bearings, drills, wood-cutting and metal-cutting tools, razors

10

Electrical and Magnetic Properties

10.1 Electron Excitation

In chapter 1 the build-up of electrons in electron shells was considered and it was stated that each electron shell and sub-shell within an atom represents a particular energy level (section 1.6). In the normal state the electrons of an atom occupy the lowest permissible energy levels. This is called the 'ground state' for the atom. As mentioned in section 1.5 it is possible for the energy of an electron to be raised to a higher level. This is called the excited state. The excited state is an unstable state and excited electrons quickly revert to their ground state and in so doing cause energy to be emitted. This energy is emitted as a discrete packet or *quantum*. These quanta released by excited atoms are called *photons* and photons travel at the speed of light. The magnitude of energy quanta is of the order of 10^{-19} J and it is customary to use the electron volt as a unit of energy for these low values, rather than the SI unit of energy, the joule.

The electron volt is the change in energy caused when an electron moves through a potential difference of one volt. The charge of one electron is 1.602×10^{-19} C, therefore one electron volt (1 eV) is equal to 1.602×10^{-19} J.

The photons emitted by an excited atom will interact with other atoms in their paths and this could result in excitation of further atoms. The photon is a means of transporting energy from one atom to another. An energy quantum may be regarded sometimes as a photon particle, but it may also be considered as a packet of electromagnetic waves of a particular frequency. The relationship between quantum energy, E, and the frequency of electromagnetic waves is Planck's quantum relationship

$$E = h\nu$$

where ν is the frequency of the energy emission and h is Planck's constant (6.625×10^{-34} J s). The concept of energy then is a dual one and depending on the circumstances, the energy quanta may be thought of as photons or waves.

Atoms may be raised from the ground state to the excited state in several ways. If crystals of common salt are placed in a flame, the flame changes to a characteristic yellow. This is because the heat energy excites the atoms making up the salt. When the electrons within the sodium atoms present revert to the ground state photons possessing energies of about 2 eV are emitted. From Planck's equation this would be equivalent to electromagnetic waves with a frequency of about 5×10^{14} Hz. This is a frequency in the yellow portion of the visible spectrum. Atoms of other elements, when heated in this way, will emit photons of different energies corresponding to waves of different frequencies. For example, strontium will give a red colouration and copper will give a green coloured flame.

Atoms may be excited in other ways than by heating in a flame. When an electric arc is struck between two electrodes the atoms in the electrode materials are excited into photon emission. Similarly when an electrical discharge is made through a vapour, as in a sodium or mercury vapour lamp, photon emission will occur.

When the atoms of an element are excited there are different levels of excitation and consequently photons of several energies are emitted (see figure 10.1). The photons of any one energy value will give rise to a radiation emission of a particular frequency but with photon emission at many discrete energy values there will be many characteristic frequencies of radiation. An excited element will therefore yield a line spectrum containing many lines characteristic of that element. This fact is used as the basis of an analytical tool and spectroscopy may be used for both qualitative and quantitative chemical analysis.

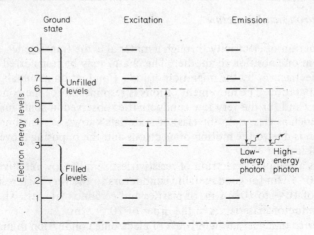

FIGURE 10.1 *Levels of electron excitation*

When a material is bombarded by a stream of high-energy electrons the degree of excitation within the atoms of the material is very great resulting in the emission of photons of very high energies. Such photons, with energies of the order of 10^2 to 10^5 eV, are X-rays with frequencies in the range 10^{17} to 10^{20} Hz. This principle is employed for the generation of X-rays.

If the energy input to an atom is large enough it is possible to excite one or more electrons sufficiently to allow them to be freed from the atom. If this occurs and an electron leaves the atom completely it leaves behind a positively charged ion. The minimum quantity of energy required to ionise an atom is termed the ionisation energy or ionisation potential of the element.

The ionisation energies of some elements of low atomic number are given in table 10.1. It will be noticed that the inert gases of Group 0 of the periodic table have high ionisation energies. It will also be noticed that elements of the second period (sodium to argon) possess lower ionisation energies than the corresponding elements of the first period (lithium to neon). The ionisation energies of the elements of any group (column) in the periodic table decrease as one moves down the column (ascending atomic number) because as the number of electron shells increases so the outer shell or valency electrons become less tightly bound to the nucleus.

TABLE 10.1

Ionisation energies of some elements (values in eV)

H							He
13.60							24.58

Li	Be	B	C	N	O	F	Ne
5.39	9.32	8.30	11.26	14.54	13.62	17.42	21.56

Na	Mg	Al	Si	P	S	Cl	Ar
5.14	7.64	5.98	8.15	10.55	10.36	13.01	15.76

10.2 Electron Band Structure

The conduction of electricity through a material is the transference of an electrical charge from one position to another. The charge may be transferred by the movement of electrons or by the migration of ions. Conduction in metals is due to electron migration. The movement of ions is responsible for the conductivity of electrolytes and for the very low conductivities observed in some insulating materials such as glasses. In the class of materials known as semiconductors conduction is due to the motion of electrons and the opposing movement of positive 'holes'.

There is a complete spectrum of resistivities from the low resistivities of the order of 10^{-8} Ωm for good metallic conductors to the very high resistivities, in the range of 10^{12} to 10^{20} Ωm, of plastics and ceramic materials. The resistivities of semiconductor materials is of the order of 10^{-3} Ωm.

In order to understand the process of electronic conduction in materials it is necessary to consider the band structure of electrons in atoms. In section 1.6 it was stated that the electrons in an atom occupy positions in electron shells, or energy levels. It was also stated that the various electron energy levels become fairly close to one another as the distance from the nucleus increases. It was also stated in section 1.10 that when atoms are in close proximity to one another the outer electron shells of adjacent atoms tend to overlap. In an aggregate of atoms the overall effect of these two factors is to cause the discrete energy levels of an individual atom to broaden into bands. This effect is illustrated in figure 10.2. The diagram shows the broadening of energy levels for the element magnesium. The dotted line in the figure corresponds to the interatomic distance in crystalline magnesium. It will be noted that at the interatomic separation distance the 3s and 3p levels overlap.

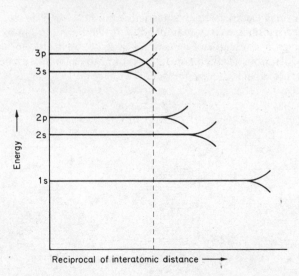

Reciprocal of interatomic distance ⟶

FIGURE 10.2 *Broadening of the energy levels of the electron states of magnesium*

When considering a large aggregate of atoms, as would be the case in a metal crystal, the energy band broadens sufficiently to contain very many energy levels, each one differing infinitesimally from the next. No two electrons can possess exactly the same energy. In the case of the element sodium (atomic number $Z = 11$) each atom contains one electron in the 3s state, even though the 3s state could hold two electrons per atom (see section 1.6).

In an aggregate of sodium atoms, if there are n atoms present there will be $2n$ available energy levels within the 3s energy band. With only one valence electron per atom, this means that the 3s band in sodium is only half filled. The valence electrons tend to occupy the lowest available levels. Energy must be given to an electron in order to move it from one position to another, that is, to conduct electricity through the material. In the case of sodium the energies of the 3s, or valence, electrons can readily be increased within the part-filled 3s band, and the metal is a good electrical conductor. In magnesium each atom possesses two 3s electrons and therefore the 3s energy band is completely filled. But in this element the 3s and 3p bands overlap and valence electrons may have their energies increased within the limits of the 3p band, and so magnesium conducts electricity readily. Adjacent permitted electron energy bands do not overlap in all materials. In diamond, for example, the four outer-shell electrons per atom completely fill the valence band and there is a large energy gap between the valence band and the next permitted energy range. The magnitude of this gap is a measure of the amount of energy an electron requires to break away from bonds and become a free electron.

In diamond, the size of this energy gap between filled valence band and the conduction band is 5.2 eV (8.33×10^{-19} J). Consequently diamond is an insulating material with a high resistivity (5×10^{12} Ωm at $20°C$). In the other group IV elements, silicon, germanium and tin, the magnitudes of the energy gaps are smaller. In silicon the energy gap is 1.1 eV (1.76×10^{-19} J), in germanium it is 0.72 eV (1.15×10^{-19} J) and in grey tin it is only 0.08 eV (0.13×10^{-19} J) (see figure 10.3). The resistivities of these elements reflect the magnitudes of the energy gaps. A small electrical field is sufficient to cause the valence electrons of tin to cross the very small energy gap into the unfilled conduction band and so tin is a reasonably good conductor of electricity. It is possible, although more difficult, for the valence electrons of silicon and germanium to cross the respective energy gaps; hence these elements are semiconductors.

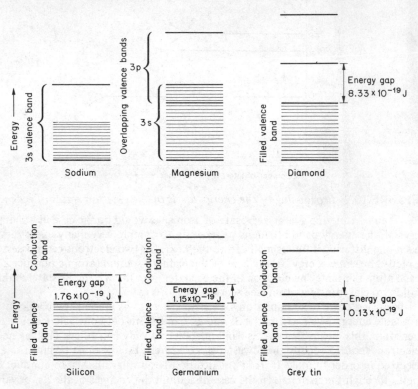

FIGURE 10.3 *Energy bands in some elements*

10.3 *Conduction in Metals*

The good electrical conductivity of a metal is due to the fact that a metallic crystal is essentially an array of positive ions in a regular pattern, with the valence electrons following random paths through the lattice (see section 1.10). Also in metals the valence electrons are either in a partially filled energy band, or adjacent energy bands overlap, as indicated in the above sections. This means that electron energies can be increased when the metal is subject to comparatively small applied electric fields. The valence electrons behave in a manner similar to free-moving gas molecules. The random-moving electrons may be in collision with atoms within the crystal lattice. As a result of a collision the electron will change its direction and may also change its velocity. Any velocity change is due to some transference of energy between an electron and an atom. When a metal is not in an electric field the algebraic sum of all the electron velocities is zero and there is no net current flow through the metal (compare this with the concept of a gas at 'rest').

When an electric field is applied there will be a drifting motion of the electrons in a particular direction, depending on the direction of the field. This causes an electrical current within the metal. The algebraic sum of all the electron velocities is not now zero but is some finite value, v, the drift velocity.

The current density, J, within the material is directly proportional to the field strength, E, and the relationship is $J = E/\rho$, where the proportionality constant, ρ, is the resistivity of the material. But the magnitude of the current density (A/m^2) is also related to the drift velocity, v, and is given by $J = nev$, where n is the number of valence electrons per unit volume and e is the charge of one electron. From these relationships the resistivity† of the material is given by $\rho = E/(nev)$. The resistivity of a metal, then, is determined by the number of valence electrons and by the drift velocity. The drift velocity is affected by the frequency of electron–atom collisions, and this will be determined by the nature of the crystal lattice and by imperfections and other variations within the lattice. Irregularities within a lattice, such as dislocations and the presence of interstitial or substitutional atoms, will increase the probability of electron–atom collisions and this will in consequence decrease the drift velocity causing an increase in resistivity (see section 8.6). When a metal is cold worked there is a considerable increase in the number of dislocations present and this is also reflected in a resistivity increase.

When a metal is heated the vibrational amplitude of the atoms within the lattice increases and this increase in the effective diameter of the atoms will again increase the probability of electron–atom collisions. In other words a temperature increase will cause an increase in the resistivity of a metal.

TABLE 10.2

Resistivities of some pure metals, alloys, and other materials

Material	Resistivity (Ωm at 20°C)	Material	Resistivity (Ωm at 20°C)
Silver	1.6×10^{-8}	Silicon	8.5×10^{-4}
Copper	1.67×10^{-8}	Germanium	10^{-3}
Aluminium	2.66×10^{-8}	Urea formaldehyde	
Magnesium	3.9×10^{-8}	(white bakelite)	10^{6}
Sodium	4.3×10^{-8}	Fireclay	1.4×10^{8}
Zinc	5.92×10^{-8}	Alumina	10^{11}
70/30 brass	6.2×10^{-8}	Phenol formaldehyde	
Nickel	6.84×10^{-8}	resins (bakelite)	10^{12}
Iron	8.85×10^{-8}	Diamond	5×10^{12}
Tin	1.15×10^{-7}	Polyethylene	10^{13}
Mild steel	1.7×10^{-7}	Nylon	10^{14}
Lead	2.1×10^{-7}	Mica	9×10^{14}
50/50 cupronickel	5.5×10^{-7}	Quartz	9×10^{14}
18/8 stainless steel	7.0×10^{-7}	Pyrex glass	10^{16}
Mercury	9.6×10^{-7}	PTFE	2×10^{16}
Graphite	1.4×10^{-5}	Vitreous silica	10^{20}

†Resistivity is a constant for a material (at constant temperature) and is expressed in units of ohm metres (Ωm). The resistance of a piece of conducting material increases with the length of the conductor and decreases with an increase in the cross sectional area. The relationship between resistance and resistivity is given by

$$\text{resistance } (\Omega) = \text{resistivity } (\Omega\text{m}) \times \frac{\text{length (m)}}{\text{c.s.a. (m}^2)}$$

When an electron gas under the influence of an electric field of strength E is moving at a drift velocity v through a metal and the individual electrons are continually in collision with the atoms and impurities of the metal there will be an energy transference from the field to the metal. The energy transfer causes an increase in the vibrations of the atoms and this corresponds to a temperature increase in the metal. The passage of an electric current through a metal causes the metal to heat up.

The effect of an increase of pressure on a metal is to decrease the volume, or in other words to increase the number of valence electrons per unit volume, with a consequent reduction in resistivity. This property may be usefully employed since it is the principle on which electrical resistance strain gauges are based.

The resistivities of a number of materials, both conducting materials and insulators, are given in table 10.2.

10.4 Semiconduction

It has already been stated that silicon and germanium are semiconductors. Their conductivity is related to the electron energy-distribution pattern within the pure material. A small number of electrons will possess sufficient energy to cross the energy gap into the conduction band, so giving the material some conductivity. Materials of this type are termed *intrinsic* semiconductors.

Silicon and germanium possess the diamond-type of crystal structure with each atom covalently bonded to four other atoms. In this type of structure the bond between any two adjacent atoms comprises a pair of shared electrons. The freeing of one valence electron to cross the energy gap into the conduction band will create a situation with only one electron being shared between one pair of atoms in the structure. This gap in the covalent bonding, shown in figure 10.4a, is called

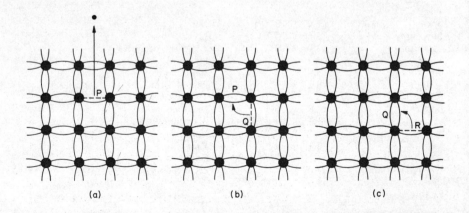

(a)   (b)   (c)

FIGURE 10.4 *Electrical conduction by movement of 'holes': (a) an electron is freed from a covalent band creating an electron hole at P; (b) electron transfer to P from adjacent site, hole effectively moves to Q; (c) electron transfer from R to Q, hole moves to R*

an *electron hole.* The freeing of any valence electron into the conduction band
generates the formation of a corresponding electron hole. The non-freed bonding
electrons are constantly in motion and an electron may switch from one pair of
atoms to fill an adjacent bonding gap. Such an electron movement will cause the
electron hole to move through the crystal lattice. This is shown diagrammatically
in figure 10.4. Since the motion of a hole is opposite in direction to the drift of
bonding electrons, the hole can be regarded as a positive charge carrier. In intrinsic
semiconductors conduction is due to both the movement of freed valence
electrons in the conduction band and the motion of positive holes. The energy
necessary to free an electron and allow it to cross the energy gap can be provided
by thermal activation. From this it can be seen that increasing the temperature of
an intrinsic semiconductor will increase the conductivity. This is the opposite of
the effect of increasing the temperature of a fully conducting metal.

Another type of semiconductor is the impurity, or *extrinsic,* semiconductor.
Silicon and germanium have four outer-shell electrons per atom and a balanced
covalently bonded crystal structure is formed. If some atoms of an element with
only three outer-shell electrons are substituted for some of the parent atoms there
will be a shortage of electrons for full covalent bonding and some electron holes
will be present in the crystal lattice. Conduction due to the migration of positive
holes can then occur. This type of semiconductor is known as *p-type.* A small
amount of aluminium dissolved in silicon or germanium is an example of this
type. Alternatively, if an element such as phosphorus with five outer-shell electrons
is introduced into the covalent silicon lattice there will be additional electrons
present. These additional electrons can be freed to enter the conduction energy
band relatively easily. This type of semiconductor, in which transference of electrical charge is by the motion of freed electrons, is termed *n-type.*

By the introduction of small controlled amounts of various elements into silicon
or germanium, semiconductors can be 'designed' to meet specific requirements.
Not all semiconductor materials are based on group IV elements. The requirement
is for an average of four outer-shell electrons. The combination of a group III and
a group V element, for example indium and antimony, will satisfy this requirement.
p-type and *n*-type semiconductors are used in combination to make rectifiers and
transistors.

10.5 The p–n Junction

A *p*-type semiconductor, although it contains positive holes due to the presence of
atoms of a group III element, is electrically neutral. Similarly, although an *n*-type
crystal contains free electrons from the group V donor atoms it is still electrically
neutral. If a *p*-type semiconductor is joined to an *n*-type semiconductor† the result
is a *p–n* junction and this is effectively a one-way valve. At the interface between
the *p*-type and *n*-type materials there will be some interaction between positive
holes and free electrons. The combination of a hole and an electron destroys both
as mobile charge carriers. The small interface zone in which the charge carriers are
lost is termed the depletion layer. This zone is very narrow, being only about

†*p–n* junctions are normally made by introducing *p*-type and *n*-type impurities into opposite
ends of a crystal of silicon or germanium, rather than by joining two separate crystals.

10^{-6} m in thickness. Within the depletion layer the p-type impurity atoms have each captured an electron and have become negatively charged ions, and the n-type impurity atoms have lost a 'free' electron to become positively charged ions. These are termed acceptor and donor atoms respectively (figure 10.5). The depletion

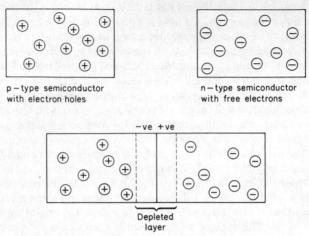

FIGURE 10.5 *p-n junction showing depletion layer*

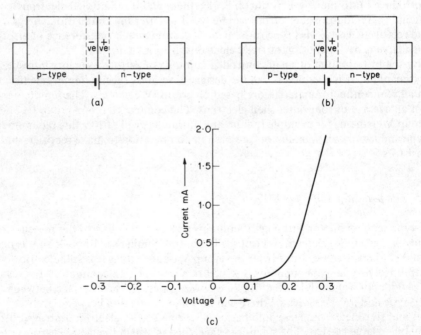

FIGURE 10.6 *p-n junction characteristics:* (a) *applied voltage tending to increase height of potential barrier;* (b) *applied voltage tending to decrease height of potential barrier;* (c) *voltage–current relationship for silicon p-n junction*

layer is thus highly charged, carrying a negative charge on one side and a positive charge on the other. This charged zone is a potential barrier and effectively prevents charge carriers from passing freely across it.

Now let us consider what happens when a voltage is applied across the ends of a crystal containing a p–n junction. If the n-type end is made positive with respect to the p-type end (figure 10.6a) the positive charge on the n-type side of the depletion layer will be increased. The height of the potential barrier at the depletion layer will thus be increased and it becomes extremely difficult for charge carriers to pass across the junction. Conversely if an applied voltage is applied in the opposite sense (figure 10.6b) the height of the potential barrier will be reduced and current will be able to flow fairly freely across the junction. Figure 10.6c showns the voltage–current relationship for a silicon p–n junction. The reverse current, with V negative, is extremely small and is of the order of one microamp.

10.6 Magnetic Domains

The motion of an electric charge creates a magnetic field. Consequently the motion of electrons within an atom produces a magnetic effect. It has been postulated that an electron not only moves in an orbital around the nucleus of an atom but also spins about its own axis, and that electron spin can be in one of two senses (to visualise this, think of clockwise and anti-clockwise spin). In most instances an element with an even number of electrons has as many electrons spinning in one sense as in the other, while an element with an odd number of electrons will just have one more electron spinning in one direction than in the other. Because of this most elements possess very weak magnetic properties and are either paramagnetic — very weakly attracted to a magnet — or diamagnetic —very slightly repelled by a magnet. Four elements, however, possess structures in which several more electrons spin in one direction than in the other and these elements are strongly attracted by a magnet. This type of behaviour is termed *ferromagnetic* and the elements that behave in this manner are iron, cobalt, nickel and gadolinium. In these ferromagnetic elements each atom behaves as a small magnet as a result of the out-of-balance electron spins.

Within a ferromagnetic material there is a sub-structure known as a *domain* structure. A magnetic domain is a small section of a crystal in which the atomic magnets are aligned parallel to one another. The dimensions of an individual domain are very small and the volume of a domain is of the order of 1×10^{-4} mm^3. In the unmagnetised condition the domains possess random orientation, but when the material is placed in a magnetic field all the domains tend to realign themselves in the direction of the field. The material is now magnetised. When the external magnetic field is removed the domain alignment may remain and the material be permanently magnetised. When a material behaves in this way it is said to be magnetically hard. A soft magnetic material, on the other hand, will lose its magnetism when the external field is removed with the domain structure reverting to a random orientation.

Whether or not a magnetic material retains its magnetism when the magnetising field is removed is largely dependent on the structure of the material. During the initial magnetisation there will be some rotation of domains and movement of domain boundaries. The movement of domain boundaries will be hindered by a strained crystal lattice, whether the lattice is strained by the presence of alloying elements or because it contains a high density of dislocations.

10.7 The Magnetisation Curve and B-H Loop

When a ferromagnetic material is magnetised the degree of magnetisation induced, B, increases as the magnetising force, H, increases until the magnetic saturation limit, B_{max}, is approached. A further increase in the magnetising force will not raise the degree of magnetisation beyond the value B_{max}. The shape of the magnetisation curve is shown in figure 10.7a.

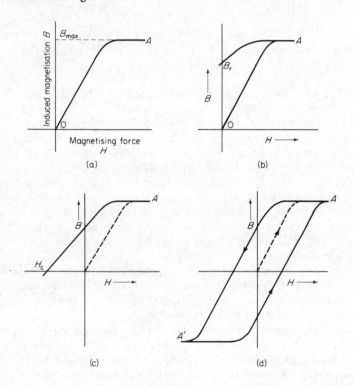

FIGURE 10.7 (a) *Magnetisation curve for ferromagnetic material;* (b) *remanence, B_r;* (c) *coercivity, H_c;* (d) *B-H loop*

If, after magnetising a material to saturation, the magnetising force is reduced, the degree of magnetisation in the material does not decrease as an exact reversal of the magnetisation curve. When the magnetising field, H, has been completely removed there will still be some magnetisation remaining in the material; this value, B_r, is called the *remanence* (see figure 10.7b). In order to reduce the degree of magnetisation to zero, the direction of the magnetising force, H, must be reversed and its value increased to some value H_c (see figure 10.7c). H_c is called the *coercivity* of the material. If the magnetising force is increased in a reverse sense beyond a value of H_c the material will be magnetised in the opposite direction until saturation is reached at A'. Reducing the magnetising field from A' and then increasing it in the positive direction will complete the magnetisation, or B–H, curve. For a material to be a good permanent magnet the remanence, B_r, should be close to the saturation value, and the coercivity, H_c, should be large. The area enclosed within the B–H loop is proportional to the energy used to realign the magnetic domains during each magnetisation cycle. This energy is then dissipated as heat. If a ferromagnetic material is used as the core of a coil through which an alternating electric current is passed it will be magnetised first in one direction and then in the other with every cycle of current. A magnetically soft material with a low remanence and low coercivity is necessary for this type of application. Figure 10.8 shows the difference in shape of B–H curves for two ferromagnetic materials.

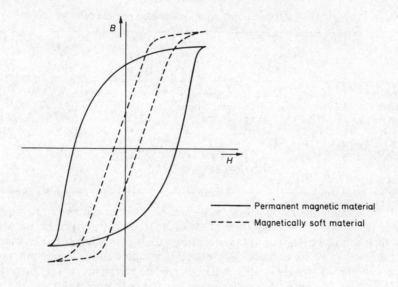

FIGURE 10.8 *B–H curves for two materials*

A ferromagnetic material that is physically soft is likely to be magnetically soft, and vice versa. In a material such as soft annealed iron the magnetic domains will have random orientation but will readily become aligned upon magnetisation. When the external magnetic field is removed, the majority of the domains will return to a state of randomness and the material will have a low remanence value. Any factor that will disturb the regularity of the iron lattice and strengthen the material physically, for example cold working or the introduction of alloying elements, will impede the rotation of domains and movement of domain boundaries and so, will increase the coercivity of the material. Highly alloyed materials containing iron, cobalt, nickel and other elements are used for the manufacture of permanent magnets. These alloys, which can be precipitation hardened to give a very highly dispersed precipitate, are most suitable because domain movement is very greatly restricted. If a magnetised material is annealed to soften it the domains can revert to a random orientation. The characteristics of some magnetic materials are given in table 10.3.

Ferromagnetic materials lose their ferromagnetism when heated above a particular temperature, the Curie temperature. The explanation for this phenomenon is that as the temperature of a material rises and the amount of thermal movement of atoms increases, a point will be reached at which the extent of atomic motion is such that the magnetic moments of the atoms become randomly orientated. The Curie temperature for pure iron is $768°C$ (1041 K) and for pure nickel is $358°C$ (631 K). The presence of alloying elements will have an effect on the Curie temperature of a metal. For example, the addition of carbon to iron will cause the Curie temperature of the material to reduce to the lower critical temperature of $723°C$ (see figure 9.12).

TABLE 10.3

Properties of some magnetic materials

Material	Use	B_r (T) (Wb/m²)	B_{sat} (T) (Wb/m²)	H_c (A/m)
Soft iron	Electromagnet cores	1.3	2.16	7
Silicon iron (3% Si)	Transformer cores	0.8	1.95	24
High carbon steel (1% C)	Permanent magnets	1.0	1.98	4.0×10^3
Fe–Co–Ni alloy (24% Co, 14% Ni, 8% Al, 4% Cu)	Permanent magnets	1.31	1.41	5.3×10^4
$NiFe_2O_4$	h.f. transformers	0.11	0.27	950
Ferrite ($MnFe_2O_4$ + $ZnFe_2O_4$)	h.f. transformers	0.14	0.36	50

In addition to metals and alloys some complex oxides can also be strongly magnetic. Magnetite, the lodestone of the ancient navigators, is magnetic. Magnetite has the formula Fe_3O_4 but of the three iron atoms per molecule one is divalent and the other two are trivalent. Magnetite is crystalline and possesses the spinel-type structure. There are other strongly magnetic spinels; these have the general formula AFe_2O_4, where A is a divalent metal and both iron atoms are trivalent. These materials are known as ferrites and, unlike metals, are insulators. This makes them suitable materials for use as magnetic cores at ultra high frequencies, since eddy currents cannot be generated within them. The *B–H* curve for a ferrite approaches a rectangular shape with an almost instantaneous saturation in the direction of the applied field.

10.8 Hysteresis and Eddy-current Losses

The terms magnetically 'hard' and magnetically 'soft' have been used to describe the behaviour of magnetic materials. For applications such as electromagnet and transformer cores, in which the core material is subjected to rapidly alternating magnetising fields, it is essential that a magnetically soft material be used. The enclosed area within the B–H loop indicates the amount of energy used to realign the magnetic domains during a complete magnetisation cycle. For a material with a coercivity of 20 A/m the hysteresis loss per cycle would be of the order of 10^{-3} J per kilogramme of material. This loss of energy is dissipated in the form of heat.

Hysteresis is not the only form of energy loss. When a conducting material is situated in an alternating field, electrical eddy currents will be generated within the material. The efficiency of a transformer depends on the total energy loss and this includes both hysteresis and eddy-current losses. The principal way in which the eddy-current losses can be reduced is by increasing the electrical resistance of the core material. The resistivity and hence the resistance of the material may be increased by alloying, and this is the effect that silicon has in iron–3 per cent silicon steel sheet for transformer cores. The resistance of the material is also increased by reducing the sectional thickness and this is one of the reasons for designing a transformer core as a laminate of thin sheet material rather than as a solid block.

10.9 Dielectrics

Insulating materials are those in which the bond structure is ionic, or covalent, or a mixture of both types. In these bonds the electrons are firmly held and are not 'free' as in metals. The energy-band diagram for an insulating material is similar to that for a semiconductor with a completely filled valence band and an energy gap between the valence band and the next permitted energy band. However, in the case of insulating materials the energy gap is so large that electrons are unable to cross it under normal conditions. An occasional electron may possess a sufficiently high energy to cross the gap, and it is this which accounts for the very high, but measurable, resistivities of insulators. At high temperatures there will be a greater probability of an occasional electron possessing the necessary energy for conduction and so the resistivities of insulators show a decrease with increasing temperature. In the case of some materials containing ionic bonds there is also the possibility that small ions in a relatively open lattice could migrate, giving some small degree of conductivity. Again, at high temperatures ions become more mobile and can diffuse more readily, reducing the measured resistivity.

Insulating materials are also known as *dielectrics*, and have the capacity to store an electrostatic charge. When a dielectric is placed in an electric field some polarisation occurs. If the material already contains polar molecules the effect is to cause the molecules to align themselves in the direction of the applied field. In non-polar materials the effect of the field is to create dipoles by causing small atomic movements leading to slight separation of positive and negative ions. When an applied electric field induces polarisation in a material there is a back effect that modifies the field. This effect is called *permittivity*. The relative permittivity, or dielectric constant, of a material is the ratio of the permittivity of the material to that of a vacuum.

The application of a stress will induce physical strain, and this may displace molecules in a polarised material, creating an electric field. The reverse is also true and variations in the external electric field will cause a change in the alignment of dipoles, creating physical strain. These effects are termed *piezoelectric*. Piezoelectric crystals are used as electromechanical transducers in ultrasonic equipment, gramophone pickups, and microphones.

Insulating materials may break down at very high voltages, the applied electric field may be sufficiently high to raise the energies of some electrons above the energy gap and thus free them. The possibility of this occurring is increased if the material contains some impurities or defects. The high-energy electrons may then disrupt covalent bonds within the insulator liberating more electrons to take part in a chain reaction. The voltage necessary to bring about this type of failure is known as the breakdown voltage. More common is surface breakdown. The presence of moisture or an accumulation of dirt on the surface of an insulator may provide a surface conduction path. To minimise the incidence of surface breakdown ceramic insulator surfaces are glazed and also made with a shaped or corrugated surface to greatly increase the surface path length.

10.10 The Hall Effect

If a current is passed through a thin slab of a conducting material and the conductor is situated in a magnetic field, such that the direction of the field is normal to both the surface of the slab and the current direction, an electric field (hence a potential difference) will be developed across the width of the conductor. This is known as the Hall effect (see figure 10.9).

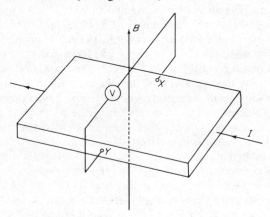

FIGURE 10.9 *The Hall effect – a potential difference is developed between X and Y*

The magnitude of the electric field, E (V/m), that is developed is given by

$$E = R_H IB$$

where I is the current density in the conductor (A/m^2), B is the magnetic field strength (Wb/m^2) and R_H is the Hall coefficient of the material. The Hall effect is observed in metals and in semiconductors. The polarity of the electric field produced in a semiconductor depends on whether the material is p-type or n-type, that is, whether the charge carriers are positive holes or electrons.

The magnitude of the Hall coefficient, R_H, is small in metals, but is very much greater in semiconductor materials 'and the Hall voltage developed across a crystal may be large enough to be put to practical use. The effect is used in a device for the measurement of magnetic-field strength. A Hall probe may be made small in size, for example 3 mm × 1 mm × 10 mm, and such an instrument may be used for the determination of magnetic-field strengths in confined spaces. One novel use of a Hall effect device is in the grading of ball- and roller-bearings. Bearings moving along an inspection line are magnetised before passing a scanning device consisting of a Hall probe, where the coercivity, H_c, is measured. There is a direct relationship between coercivity and the physical hardness of the bearing. The measuring device feeds information to an electronic sorting-gate so that all bearings with coercivities outside the acceptable limits for the product are separated from the rest. The satisfactory bearings are demagnetised before use. A schematic diagram of such an inspection line is shown in figure 10.10.

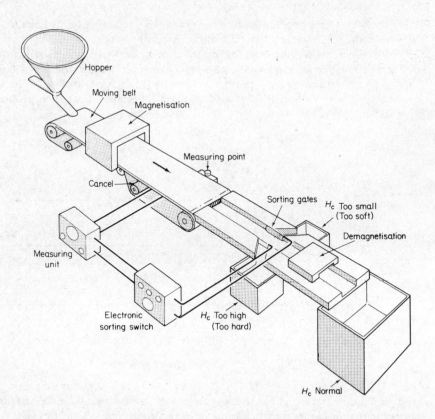

FIGURE 10.10 *Inspection line using a Hall device*

11

Optical, Thermal and Other Properties

11.1 The Energy Spectrum

Energy may be transmitted in the form of a wave-like radiation termed electro-magnetic radiation. Electromagnetic radiations may possess an extremely wide range of frequencies ranging from low radio-frequencies, through infra-red, visible light and ultra-violet into the X-ray and γ-ray range. Electromagnetic waves travel through a vacuum at a velocity of $c = 3 \times 10^8$ m/s (c = velocity of light). The frequency, ν, and wavelength, λ, of any radiation are related, since their product is a constant

$$\nu\lambda = c$$

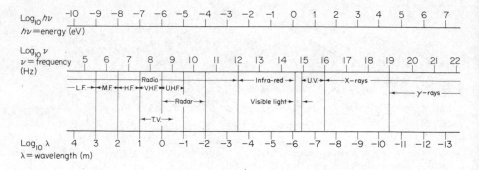

FIGURE 11.1 *Electromagnetic spectrum*

As stated in section 10.1 electromagnetic waves can be regarded as a series of energy quanta, or photons. Planck's quantum relationship $E = h\nu$ relates the energies of photons to the frequency of the radiation; h is Planck's constant $(6.625 \times 10^{-34}$ J s).

The energy of a photon for a low-frequency radio wave of frequency 200 kHz is very small and has a value of about 13×10^{-29} J (8×10^{-10} eV). The energy of a photon of light in the visible sector of the spectrum, frequences of almost 10^{15} Hz, is about 2 eV while for X- and γ-radiation with frequencies in excess of 10^{16} Hz, photon energies will range between 100 and 10^6 eV. High-energy photons may cause changes in a material when the material is subject to the radiation (see section 11.13).

11.2 Absorption and Transmission

In metals the outer-shell, or valency, electrons are in partially filled conduction bands (see section 10.2). When photons of comparatively low energy are incident upon a metal the photon energy can be transferred to valency electrons, raising the energies of these electrons within the conduction band. When this occurs the incident energy is absorbed by the material and not transmitted through it. Metals are opaque to radio waves, infra-red, visible light and ultra-violet but the high-energy photons of X- and γ-rays can penetrate through the metals. There is some absorption of this high-frequency radiation and the total amount of X- or γ-radiation that is transmitted through the metal is inversely proportional to the thickness. The absorptive power of a metal for X-radiation is also in proportion to the density of the metal. X- and γ-radiation are used for the radiography of metals and greater penetration is achieved by the harder (higher-frequency) radiation. There is some reflection of X-rays from atomic nuclei and because of this fact X-rays can be used for investigation of crystal structures.

Polished metal surfaces will largely reflect, rather than absorb incident low-energy photons. This reflection is partially selective. For example, copper absorbs visible light at the short wavelength end of the visible spectrum, hence its reddish colour.

Dielectric materials possess filled electron-energy bands and forbidden energy gaps. There cannot be transference of energy from photons to electrons in these materials in the same way as in metals and so these materials generally do not absorb electromagnetic radiation but transmit it.† Good radio reception is obtained within a building since the brick and glass are largely transparent to radio waves, but radio reception within a metal-bodied car is difficult without an external aerial.

The explanation of absorption and transmission given above is a very simple one and does not cover the situation fully. Non-metallic materials are not fully transparent to all types of electromagnetic radiation and many materials show absorption peaks at particular frequencies. For example, window glass is opaque to the infra-red section of the spectrum. Also certain materials show definite colours because they strongly absorb certain frequencies within the visible part of the energy spectrum.

Semiconductor materials possess filled valence bands but with forbidden energy gaps of a much smaller size than those in insulators. Low-frequency radiation is not absorbed by semiconductors but if the frequency of the incident radiation is such that the photon energy is equal to the magnitude of the energy gap, energy may be transferred from photons to electrons. In other words, the incident energy

†Most non-metallic materials are transparent to light in thin section. Their opaqueness in thick section is largely due to repeated reflection and refraction at internal surfaces.

is absorbed and electrons are excited into the conduction band. This gives rise to the effect termed *photo-conduction* since the conductivity of the material increases in a similar manner to that caused by raising the temperature of the semiconductor.

11.3 Refraction and Polarisation

A transparent material possesses a refractive index, μ, and this is defined as being the ratio of the velocity of light in a vacuum, c, to the velocity of light in the material, v.

$$\mu = \frac{c}{v}$$

There is a relationship between the refractive index of a material and the dielectric constant, K, for the material

$$\mu^2 = K \quad \text{or} \quad \mu = \sqrt{K}$$

The velocity of light and the refractive index of a material are functions of frequency. An example of this is the separation of white light into the colour spectrum by a glass prism; this effect is called *dispersion*.

There are very many different types of glass available for optical purposes and they possess refractive indices in the range 1.5 to 1.7. A material with a refractive index of this order of magnitude is ideal for use in lenses and similar devices. Refractive index is not the only important property, and an optical material must also possess a high transmission value. The transmission ability of optical glass is about 98 per cent (that is, 98 per cent of incident light would be transmitted through a section of 10 mm thickness). A number of plastics materials are also suitable for optical applications, possessing similar refractive indices to glass and with transmission abilities of 90 per cent or more.

A glass is isotropic, but crystalline materials are not. If crystallisation takes place in a glass or glassy polymer the crystals will possess a different refractive index from the glass. Reflections at internal interfaces between glass and crystal regions will lead to the material becoming translucent.

Due to the anisotropy of crystals light will have a different velocity in the directions of the different crystal axes (except in cubic crystals where the velocity of light is the same for each major axis). Crystals in which the velocity, and hence the refractive index, differs from one crystal axis to another are termed *birefringent*. Birefringent crystals are doubly refracting. A beam of light incident upon the crystal is split into two beams. There is another interesting effect in that the two beams are polarised at right angles to each other. Some crystals have the effect of absorbing one polarised component of the split beam to a very much greater extent than the other resulting in the emergent beam being of polarised light. This is the effect when light passes through Polaroid.

An isotropic material, such as glassy polystyrene, becomes anisotropic to some extent when it is physically strained. When physically strained the material becomes birefringent. This stress-induced birefringence is termed the *photoelastic* effect. In photoelastic stress-analysis a model of a component, say a chain hook, is made from clear plastic. The model is loaded and, due to the birefringence, interference fringes can be observed within the model. The fringe pattern will indicate the stress distribution within the model and the number of fringes observed is in direct proportion to the magnitude of the stress induced.

11.4 Luminescence

When relatively high-energy photons are incident on a material some of the valency electrons may be excited to a higher energy band, as mentioned in section 10.1. When the excited electrons drop back to the ground state, or some intermediate energy level, energy will be emitted from the material. The emission is often at a lower frequency than the incident radiation. When the emission is in the visible portion of the spectrum the general phenomenon is termed *luminescence*. When the emission takes place immediately it is termed *fluorescence*. Many materials fluoresce in ultra-violet light and appear to be of a different colour from that observed in white light. *Phosphorescent* materials emit light after a short delay, and the emission usually lasts for a short time period. Phosphorescent materials are used to coat the surface of television and radar tubes. For the display tubes of search radar the phosphorescent coating is selected to give an after-glow period of several seconds.

11.5 Thermal Conductivity

If one end of a bar of material is heated the atoms in that portion will be stimulated into increased vibration. Energy will then be transmitted along the bar from the hot to the cool end in an attempt to reduce the energy at the heated end. For an electrically insulating material the energy transference along the bar will be a result of atomic vibrations only. The transmission of heat energy along the bar will be a function of the separation distance between atoms, the density of the material and the specific heat. The thermal conductivity of a glass or liquid will be less than the conductivity of the crystalline version of the same material, since the atoms in a glass or liquid are not in a fully ordered array and interatomic separation distances are variable.

In metals, with valence electrons existing as a free cloud or gas, energy may be transmitted along a bar through the medium of electron movement as well as by means of atomic vibration. Metals are, in general, good conductors of both electricity and heat. The class of materials known as electrical semiconductors are also reasonably good thermal conductors because heating the material can cause some electrons to be excited and cross the energy gap into the conduction band. The excited electrons may then take part in the transmission of heat energy from one part of the material to another.

The thermal conductivities of some materials are listed in table 11.1. The units of thermal conductivity are watts per metre per degree Kelvin (W/m K).

It will be noticed that the thermal conductivities of gases, apart from hydrogen, are very much lower than the conductivities of other non-metallic materials. Hence the reason for most thermal insulating materials being porous. Cavity-wall construction and double glazing in buildings are other examples in which use is made of the insulating properties of gases. There is no thermal conduction at all in a vacuum, since there are no atoms present to vibrate and transmit energy. This principle is used in the vacuum flask in which there is an evacuated space between two layers of glass.

The rate of heat transmission through a material is given by the expression $kA (T_1 - T_2)/t$ where k is the thermal conductivity, A is the surface area of the material, T_1 and T_2 are the temperatures of each surface of the material and t is the thickness.

TABLE 11.1

Thermal conductivities of some materials at 20°C

Material	Conductivity (W/m K)	Material	Conductivity (W/m K)
Silver	420	Building brick	0.63
Copper	395	Polyethylene	0.33
Aluminium	240	Nylon	0.25
Magnesium	168	Paraffin wax	0.25
Nickel	88	Phenol-formaldehyde	0.17
Mild steel	51	Rubber	0.13
Lead	33.5	Polystyrene	0.08
Quartz	5.1	Sand	0.042
Silica glass	1.25	Hydrogen†	0.17
Concrete	1.05	Oxygen†	0.025
Mica	0.84	Nitrogen†	0.024
Plate glass	0.75	Perfect vacuum	zero

†At atmospheric pressure

Example　　Determine the heat loss through a brick wall 4 m by 3 m of 0.25 m thickness if the inner surface is maintained at 20°C and the outer surface temperature is 5°C. The thermal conductivity of brick is 0.5 W/m K.

$$\text{Heat loss} = \frac{k A (T_1 - T_2)}{t}$$

$$= \frac{0.5 \times 4 \times 3 \times 15}{0.25}$$

$$= 360 \text{ W}$$

11.6 Thermal Expansion

When a material is heated the vibrations of its constituent atoms increase. This is accompanied by an increase in the physical dimensions of the material, that is, expansion. If the material is constrained during heating or cooling so that dimensional change is restricted, thermal stresses can be set up within it (see section 4.7). The rapid heating or cooling of non-ductile materials such as ceramics and glass can cause cracking.

The change in length, δl, that occurs when a material of length l_0 is heated from temperature T_0 to T_1 is given by

$$\delta l = l_0\, \alpha\, (T_1 - T_0)$$

where α is the linear expansion coefficient for the material. The value of the expansion coefficient is very small for ceramic materials. The values of α for metals is about twice that for most ceramics, while plastics materials have very high thermal expansion coefficients. Table 11.2 lists the thermal expansion values for some materials.

TABLE 11.2

Thermal expansion coefficients for some materials

Material	α ($/^\circ$C)	Material	α ($/^\circ$C)
Silica glass	0.54×10^{-6}	Aluminium	21.6×10^{-6}
Graphite	5.4×10^{-6}	Magnesium	25.2×10^{-6}
Plate glass	9.0×10^{-6}	Polystyrene	63.0×10^{-6}
Building brick	9.0×10^{-6}	Phenol-formaldehyde	72.0×10^{-6}
Mild steel	11.7×10^{-6}	Rubber	80.0×10^{-6}
Concrete	12.6×10^{-6}	Nylon	100.0×10^{-6}
Copper	16.2×10^{-6}	Polyethylene	180.0×10^{-6}

When a material melts there is generally a marked increase in volume accompanying the transition from solid to liquid. In the case of metals the expansion at the melting point is of the order of 6 per cent. There are certain exceptions to this, a notable one being antimony, which contracts on melting. Use is made of this peculiarity in the manufacture of type metals. These contain some antimony in order to prevent solidification shrinkage and ensuring that a good clean type face is obtained. Ice is another solid that contracts on melting, with the reverse process, expansion during freezing, giving rise to burst water-pipes.

11.7 Sound Absorption and Damping

Sound energy is transmitted through a medium both as a compression wave and as a shear wave. Sound cannot travel through a vacuum because there are no atoms available to vibrate and transmit energy. The velocity of sound within a material is related to both the density and the elastic moduli. The velocity, v_l, of a longitudinal compression wave within a material is given by

$$v_l = \sqrt{\frac{K + \left(\frac{4}{3}\right)G}{\rho}}$$

where K is the bulk modulus of elasticity, G is the modulus of rigidity, and ρ is the density of the material. The shear wave velocity, v_s, is given by

$$v_s = \sqrt{\frac{G}{\rho}}$$

The elastic constants of a material can be determined from a knowledge of the velocity of sound through it. All the elastic constants are related to one another (see section 4.5) and so values for the modulus of elasticity, E, and Poisson's ratio, ν, can also be determined in this way.

When sound travels through a medium some of the energy is absorbed by the material. This absorption, or damping of vibrations, is due to the internal friction of the material. The internal friction of a material is a function of the degree of crystallinity and the nature of the interatomic bonds within the substance, but it is also greatly affected by the presence of discontinuities within the material, such as impurities, and cracks. These serve to increase the internal friction. The internal friction of metals, and a number of non-metallic crystalline solids, is generally low. The internal friction of brick and concrete is about 100 times greater than that of many metals, while that of timber is about 1000 times greater than that of many metals. Most thermoplastic materials have internal friction values similar to or greater than those for timber.

For sound-insulation purposes the most effective type of material is one of low density and high internal friction. Foamed plastics are an excellent example of this. When sound energy reaches an interface between two different media there will be some reflection and this reflection may approach 100 per cent at an air–dense-solid interface. It is for this reason that double glazing is such an effective sound insulator since there are four air–glass interfaces between the external noise source and the ear, with a low density medium, air, between the two glass layers.

A material of high internal friction will have great ability to attenuate vibrations, and vice versa. This ability is termed the damping capacity of the material. The damping capacity, ζ, is given by

$$\zeta = \frac{1}{n}$$

where n is the number of vibrations in free attenuation from an amplitude A to an amplitude A/e, that is, $0.368\,A$. For the manufacture of a bell a material of very low damping capacity is necessary, hence the use of the copper alloys, brass and bronze. Cast iron is a metal with a very high damping capacity because of the presence of graphite flakes and small flaws in its structure. It is therefore a most suitable material for components where vibrations should be absent, as in a lathe bed. For antivibration mountings, very high internal friction materials such as rubber and cork are used.

11.8 Ultrasonics

The upper limit of sound frequency audible to the human ear is about 15 kHz. Above this frequency is the ultrasonic range. Ultrasound is used for many purposes, including underwater 'sonar' and the non-destructive examination of materials. Ultrasonic techniques are highly suitable for the detection of sub-surface defects in materials. Surface defects may also be detected using ultrasonic testing, but other methods are generally more satisfactory. For metals inspection, frequencies in the range from 0.5 to 15 MHz are generally used. When an ultrasonic wave passes from one medium to another, some reflection takes place (this reflection can approach 100 per cent at a metal–air interface). Any defect such as an internal crack, porosity, or a non-metallic inclusion will therefore act as a reflecting surface for ultrasonic waves. Ultrasonic waves are produced by stimulating a piezoelectric crystal into high-frequency vibration by an electrical impulse. The crystal is highly damped so that only a short-duration pulse is created, and the crystal is then dormant and in a position to receive an echo and convert the ultrasonic echo into an electrical signal. Quartz, barium titanate, and lead niobate are typical piezoelectric materials used in the manufacture of ultrasonic probes. Because ultrasound is quickly dissipated in air, there must be close coupling through a high-density medium between the crystal face and the surface of the metal under test. This is achieved by having a film of oil or water on the metal surface. The time of pulse transmission, and the reflected echo signals are displayed on a cathode-ray tube. Ultrasonic inspection is an extremely versatile system and may be used for practically any material, and it can also be used for testing large thicknesses of material, for example steel thicknesses of up to 15 m can be satisfactorily tested.

11.9 Interface Effects and Surface Tension

The properties at a surface are different from those within the heart of a material. An atom or molecule in the centre of a liquid is completely surrounded by other molecules and it is attracted equally by all its neighbours. A molecule at the surface, however, will be attracted inwards by the molecules within the bulk material. The net result of this inward pull is that the surface of the liquid tends to contract to the smallest possible area. It is for this reason that liquid droplets and gas bubbles within liquids tend to be spherical. The surface behaves as if it is in a state of tension, and this phenomenon is termed *surface tension*. The term signifies a separation between two substances, for example liquid–gas, and the value of surface tension is dependent on both substances present. The recorded values of surface tensions for liquids are generally for a liquid–air interface. The surface-tension value may well be different for the interface between the liquid and some other gas. The interface surface-tension, γ, has the units of newtons per metre (N/m) and the values of γ for some liquids are given in table 11.3.

TABLE 11.3
Surface tensions of some liquids in air (at 20°C)

Liquid	γ (N/m)	Liquid	γ (N/m)
Ethyl alcohol	0.022	Benzene	0.029
Acetone	0.024	Water	0.073
Carbon tetrachloride	0.027	Mercury	0.485

It will be noticed that the surface tension of water is high compared with organic liquids, even though it is very low compared with mercury.

Whether a liquid wets the surface of a solid or not depends on the values of interface surface-tensions for the three interfaces involved, namely liquid–air (γ_{la}), liquid–solid (γ_{ls}) and air–solid (γ_{as}). Consider a liquid resting on the surface of a solid. At equilibrium there will be an angle of contact, θ. The angle θ is measured within the liquid (see figure 11.2). At equilibrium, the interface forces will be in balance, and this can be represented by the statement

$$\gamma_{as} = \gamma_{ls} + \gamma_{la} \cos \theta$$

If $\gamma_{as} > \gamma_{ls}$ then $\cos \theta$ is positive and $\theta < 90°$; if $\gamma_{as} < \gamma_{ls}$ then $\cos \theta$ is negative and $\theta > 90°$. When the contact angle, θ, is less than $90°$ the liquid is said to wet the solid surface. This is the case with water on glass. If the contact angle, θ, is greater than $90°$, as with mercury on glass, the liquid does not wet the solid surface.

The air–liquid interface for liquid in a tube will be curved. When the liquid wetting liquids the surface will be convex upwards. The curved liquid surface is termed *meniscus* and readers are probably familiar with meniscus curves for both water and mercury in glass tubes.

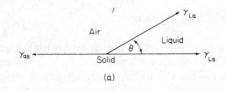

(a)

FIGURE 11.2   (a) *Liquid–solid contact angle;* (b) *liquid wetting solid;* (c) *non-wetting liquid*

The pressure of liquid close to a curved surface is different from that close to a plane surface by an amount, δP, given by

$$\delta P = \frac{2\gamma}{r}$$

where r is the radius of curvature. When the liquid surface is convex δP is positive and when the surface is concave δP is negative. When an open-ended fine-bore, or capillary, tube is partially immersed in water the meniscus within the tube will be concave with a small radius r, compared with the flat surface of water outside the tube. The pressure just below the meniscus in the tube will be less than the pressure just below the plane water-surface in the main vessel by an amount $2\gamma/r$ and so the water level within the tube will rise through some height h, where h is the height of a column of water having a pressure of $2\gamma/r$ at its base. (The level of a liquid with an angle of contact, θ, greater than $90°$, for example mercury, will be depressed.) This phenomenon is termed *capillary attraction* or *capillarity*. In general, liquids of low surface tension will have low contact angles and show high capillarity.

The surface of a solid–gas interface is in a state of tension in a similar manner to a liquid–gas surface. In a strongly bonded crystalline solid the atoms in the surface layers have their bonds only partially satisfied and they can react or join with atoms from the other medium, providing an adsorbed film on the surface of the solid. In some cases the forces of attraction between atoms on the solid and the adsorbed layer are of the van der Waals type but in other cases there may be ionic or covalent type bonding between the solid surface and atoms from the surrounding medium. This type of effect occurs between clean metal surfaces and oxygen and is an important first stage in the surface oxidation of metals. When there are ionic or covalent bonds formed at the surface the phenomenon is called chemisorption.

11.10 Friction and Lubrication

The interface between two solids is of considerable importance in engineering. The surface of a solid is normally far from smooth when examined on a microscale. A lathe-turned or flat-ground surface will possess innumerable 'hills and valleys'. The height difference between hills and valleys for a fine-ground surface is of the order of 1–3 μm. When two metal surfaces are put together the true contact area, because of surface irregularities, will be very much less than the apparent area of contact. In consequence the stresses involved at points of contact may be sufficiently high to cause localised deformation and possibly welding. If a shearing force is applied in such a way as to cause sliding between the contacting surfaces, sliding motion will be resisted. The resistance to sliding is termed friction and is due to both the roughness of the surfaces and localised welding that may have occurred at points of contact.

When dry solid surfaces are in sliding contact the work done in overcoming friction is converted to heat energy and there will be considerable abrasion between them. This leads to severe wear of the surfaces. The surface temperature build-up could be sufficient to cause the two surfaces to weld over an extensive area. This is termed *seizure*. The extent of the friction between surfaces is considerably reduced by the presence of a lubricant, that is, a fluid layer interposed between the two solid surfaces.

In normal fluid lubrication, also termed hydrodynamic lubrication, the two solid surfaces are kept apart by means of a fluid layer and provided that the separation distance between the surfaces does not fall below about 0.1 μm almost any fluid that does not react with the solids may be used. Most lubricating fluids are mineral oils but high-pressure air, as a fluid, is used in low friction air bearings. Most lubricating oils contain boundary-layer additives. Such an additive is composed of polarised molecules which become strongly adsorbed on a metal surface. When conditions are such that hydrodynamic lubrication may break down, the adsorbed layers will still keep the solid surfaces apart. Organic fatty acids are used as boundary-layer additives in oils, as is also molybdenum disulphide.

11.11 Chemical Attack and Corrosion

Thermoplastic and thermosetting materials are, in general, highly resistant to chemical attack. Most plastics are largely unaffected by weak acids and alkalis, although they may be affected by concentrated reagents. They are also highly resistant to many oils and greases, but may be dissolved by some organic solvents. Plastics materials are also unaffected by microbial organisms. The resistance of plastics materials to reagents and organisms poses a major problem from the point of view of waste disposal, and research is at present taking place into the possibilities of microbial methods for the breakdown of plastic refuse.

Timbers are subject to bacterial and fungal attack, and numerous organisms feed on moist wood fibre. Thoroughly seasoned timber is not such good food value to the bacteria as moist wood, and the rate of attack is very greatly reduced in dry wood. There are also larger organisms that feed on certain types of wood. The teredo is a small marine animal that attacks timber immersed in sea water, while terrestrial insects, such as wood worm and, in some parts of the world, termites, can also be a problem. Preservatives may be used to retard the decay of timber. A common and cheap preservative is creosote, since this is poisonous to the wood-attacking organisms. There are many other wood preservatives available that contain specific fungicides.

Metals react with oxygen to form oxides and this type of reaction will occur at any clean metal surface in contact with air. In some instances the oxide layer formed at the metal surface is impervious to oxygen and serves as a thin protective layer. This is the case with metals such as aluminium and chromium. Some metallic oxides are porous and oxidation may continue, but even with these metals direct oxidation only assumes importance at elevated temperatures when oxidation rates become appreciable.

The corrosion of metals is largely electrochemical and this subject was introduced in sections 2.9 and 2.10 and the standard electrode potentials of some metals were quoted in table 2.2. The order in which the various metals are placed in table 2.2 is called the *electrochemical series.*

It should be noted, however, that the potential differences between metals in equilibrium with their ions may differ from the standard values if the metals are not immersed in a standard 'normal' electrolyte but in some other electrolyte, for example, sea water. Also, the order of metals in the series may differ from the order of the standard electrochemical series when metals are immersed in other electrolytes. The order of metals, when determined for some electrolyte other than a normal solution of hydrogen ions is termed a *galvanic* series.

In any galvanic cell, it is always the anode that will be corroded, while the cathode will be protected against corrosion. Galvanic, or corrosion, cells may be established in several ways, and the main types of galvanic cells are (a) composition cells, (b) stress cells, and (c) concentration cells. An obvious example of a composition cell is two dissimilar metals that are in direct electrical contact and also joined via a common electrolyte. For this reason, dissimilar metals should not be in contact in situations where corrosion conditions may prevail. It is also possible to have composition cells formed within one piece of material. In any two-phase alloy, or a metal containing impurities, one phase may be anodic with respect to the other and galvanic microcells may be established. Examples of this are pearlite in steels, and precipitate particles of chromium carbide in austenitic stainless steels. In pearlite, the ferrite is anodic with respect to cementite, and in austenitic stainless steels the grain-boundary precipitate of chromium carbide, which is formed if the steel is incorrectly heat treated, is anodic with reference to the austenite solid solution. This latter fact is responsible for the 'weld decay' type of corrosion. From the above, it follows that normally pure metals or single-phase alloys will possess better resistance to corrosion than impure metals or multi-phase alloys.

Atoms within a highly stressed metal will tend to ionise to a greater extent than atoms of the same metal in an annealed condition. Consequently the stressed material will be anodic with respect to the unstressed metal. A stress-type of galvanic cell may be established in a component, or in a structure, where the stress distribution is uneven; (this also applies to an uneven distribution of residual stresses within a cold-worked metal). Some alloys are particularly prone to stress-corrosion failure, which is a grain-boundary (inter-crystalline) type of corrosion. This may occur in cold worked α-brasses, if they are not properly stress relieved. Randomly arranged grain boundary atoms tend to ionise more rapidly than atoms within a regular crystal lattice, and so a grain boundary tends to be anodic with respect to a crystal grain. Much corrosion is of an inter-crystalline nature. It also means that a coarse-grained metal tends to possess a better resistance to corrosion than a fine-grained sample of the same metal.

When a metal is in contact with a concentrated electrolyte solution it will not ionise to as great an extent as when it is in contact with a dilute electrolyte. In other words, if one piece of metal is in contact with an electrolyte of varying concentration, those portions in contact with dilute electrolyte will be anodic with reference to portions in contact with more concentrated electrolyte. Galvanic cells of the concentration type may be encountered in situations involving flowing electrolytes in pipes and ducts.

11.12 Corrosion Prevention

It is almost impossible to prevent corrosion completely, but steps may be taken to minimise the problem. One of the most common general techniques is to isolate the material from its environment. There are very many types of protective coatings used for metals, including paint, vitreous enamel, plastic, a layer of another metal and oxide coatings. Painting is a cheap and convenient method for protecting metal surfaces from corrosion. It does, however, possess several disadvantages. A paint layer does not possess high wear resistance, it deteriorates with time and surfaces will require periodic repainting, and paint cannot be used for the protection of components that are to operate at elevated temperatures. Vitreous enamel coatings are very good for many purposes and they will be suitable for use at elevated temperatures, for example, vitreous enamelled steel is used for the manufacture of gas and electric ovens. The major disadvantage of this type of coating is its brittle nature. Steel is sometimes coated with a thermoplastic. This type of coating is widely used in the manufacture of welded wire trays for refrigerators and deep freezers.

It is often convenient to protect one metal from corrosion by covering it with a thin layer of a metal with a good corrosion resistance. Aluminium, cadmium, chromium, nickel, tin, and zinc are used for coating steels. The coating metal may either be anodic or cathodic with respect to the underlying steel. The type of coating used will have a significant effect on subsequent corrosion if the coating layer is imperfect, or becomes broken during service. With a cathodic coating, such as tin, on steel, any break in the coating will establish a galvanic cell with the steel acting as the anode. In this case the rate of corrosion of exposed steel will be rapid. On the other hand, if there is a break in a coating of an anodic metal, such as zinc on steel, galvanic corrosion of the remainder of the surface coating will occur, and some protection will continue to be offered to the steel until most of the coating metal has been removed.

In some instances a metal surface may become *passified* and the corrosion rate reduced. If strong oxidising conditions exist within an electrolyte an anode reaction may occur to form an oxide layer on the surface of the metal. Such an oxide layer may be protective. Stainless steels are highly resistant to corrosion by virtue of the passive layer of chromic oxide that forms on the metal surface, but in reducing conditions the oxide layer may break down and the material is no longer protected. For the protection of ordinary steels and irons in closed systems, such as radiators and boilers, *inhibitors* may be added to the water. These are generally compounds that will provide chromate ions or phosphate ions, and these help to maintain passive films on the steel surface.

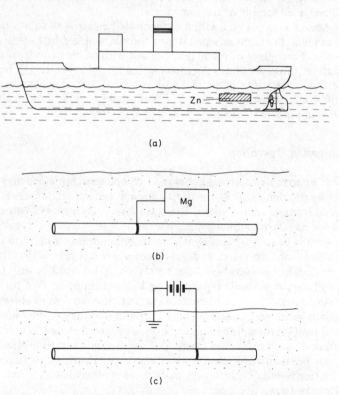

(a)

(b)

(c)

FIGURE 11.3 *Cathodic protection:* (a) *protection of ship's hull below waterline;* (b) *protection of buried pipe-line with magnesium anode;* (c) *protection of buried pipe-line using impressed voltage*

Another method used for the prevention of corrosion is *cathodic protection.* In this method, galvanic cells are deliberately created and the material, or structure, that is to be protected is made the cathode. Zinc, aluminium, and magnesium are all metals that are strongly anodic with respect to steel, and these metals may be used as sacrificial anodes to prevent the corrosion of buried steel structures or pipework, or for the protection of ships' hulls below the waterline. It is easy and relatively inexpensive to replace anodes when necessary. The function of a sacrificial anode is to maintain a supply of excess electrons to the material that is to be protected. This may also be achieved by applying an impressed voltage to the structure requiring protection (see figure 11.3).

11.13 Irradiation of Materials

Radiation may have an effect on materials. Electromagnetic radiation of various frequency ranges will have an effect on some materials. A good example of this is the effect of radiation in the infra-red, visible light, ultra-violet, and X-ray ranges on photographic emulsions. The radiation causes dissociation of silver from the silver salts in the emulsion. Similarly, visible light, ultra-violet light, or X- and γ-radiation will stimulate numerous chemical processes. An example is the effect of X- or γ-radiation on a linear polymer, such as polyethylene. The radiation increases the energy of the polymer and promotes branching and cross-linking reactions, thus producing thermoplastics of greater rigidity. This process is utilised in the production of some plastic components. Bombardment of a thermoplastic with electrons (β-rays) will also produce similar effects. The effect of high frequency electromagnetic radiation on some compounds is to excite electrons to higher states of energy. When the excited electrons subsequently fall back to lower energy states, energy is emitted, often in the visible light range, producing fluorescence and phosphorescence (see section 11.4). A similar effect is noticed when some compounds are subjected to β-radiation (electrons), and this is the principle used in cathode-ray tube coatings. High-frequency radiation or electron bombardment will not affect metals, but the radiation will be partially absorbed by the material. A considerable amount of radiation will pass through the metal, and this principle is utilised in the radiography of metals and in electron-transmission microscopy. Because the wavelengths of X- and γ-radiation approach the sizes of individual atoms, there will also be diffraction effects, and diffraction patterns are used in the identification and analysis of crystals.

The irradiation of a material with neutrons is invariably damaging. A neutron is a small particle possessing no electrical charge and a fast neutron will not be attracted to or deflected by charged particles, such as electrons and protons. A fast neutron may travel through a large number of atoms before it collides with an atomic nucleus. When such a collision occurs the atom, or ion, that has been hit is displaced from its normal position, and the path of the neutron will be deflected. A neutron will lose some energy at each collision and, eventually, a lower energy neutron will be 'captured' by an atomic nucleus. When a nucleus captures a neutron, the nucleus will become unstable and will emit α-, β-, or γ-radiation, and in so doing may be changed into a radioactive isotope of another element.

The irradiation of polymers with neutrons at low neutron flux densities will cause some cross-linking and branching to occur, in a similar manner to activation by X-rays, but at higher neutron flux densities the effect of neutron collisons is to cause degradation of polymers. The effect of a fast neutron colliding with the nucleus of an atom in a metallic crystal lattice is to cause the atom to be knocked out of its equilibrium position within the lattice. The knocked-out atom moves into an interstitial lattice position, leaving behind a vacancy. The lattice defects created by neutron collisions make the material harder and increase the yield strength. They also make the material more brittle. In the case of those metals that show ductile–brittle transition behaviour, the transition temperature is increased; (the neutron irradiation of steels will raise ductile–brittle transition temperatures by up to 100°C). This type of radiation damage may be removed by annealing the material. Irradiation with slow neutrons will give a different type of radiation damage. Slow neutrons may be captured by atomic nuclei and the irradiated metal will become radioactive or 'hot'. Also the transmution product may alloy with or react chemically with the other atoms of the metal.

12

Forming Processes

12.1 Introduction

The finished engineering component or any form of consumer hardware may be required in any one of an innumerable number of shapes and sizes. A finished component may range from a very large casting of many tonnes mass down to a minute component in the electronics field. There are very many types of forming process that may be used for the production of shapes and some of these will be discussed in this book. The total range of shaping processes can be broadly grouped into four major categories

(1) casting, namely the pouring of liquid into prepared moulds,
(2) manipulative processes involving plastic deformation of the material,
(3) powder techniques, in which a shape is produced by the compaction of a powder, and
(4) cutting and grinding operations.

The physical and other properties of a material help to determine the type of process most suitable for shaping it. The following examples illustrate this statement.

(1) Zinc cannot easily be deformed plastically because of its hexagonal crystal structure, but it is very suitable for diecasting owing to its relatively low melting temperature (419°C).
(2) Polystyrene softens and becomes very plastic at temperatures just above 100°C and lends itself to forming by a plastic-deformation process at these comparatively low temperatures.

But while the properties of a material influence the type of forming process that may be used, so does the type of processing affect the properties of the material in the final article. For example a metal cast in a sand mould will possess a larger crystal grain size than a similar mass of the same metal cast into a metal chill mould. This will make the sand casting somewhat less strong than the chill casting. Both castings may contain some porosity whereas a forging of the same metal will be free from porosity making the forging stronger than either casting.

12.2 Sand Casting

The casting of liquid metal into a shaped mould and allowing it to solidify is a very convenient way of making solid metal components. One of the oldest casting techniques is sand casting. A mould is made by ramming moulding sand (basically a silica sand with a proportion of clay as a binding agent) around a pattern of the part to be made. The pattern, which is generally made of hard wood, has to be made somewhat larger than the required dimensions of the finished casting, to allow for contraction of the casting during cooling. The mould is made in two or more parts, to facilitate removal of the pattern, and feeder channels, gates, and risers must also be incorporated in the mould (see figure 12.1a). Hollow castings may be made by fitting cores in the mould. Cores, which have to be strong enough to be handled and also to be able to remain largely unsupported within the mould, are often made from sand bonded with linseed oil, or made by the shell-moulding process from sand–resin mixes.

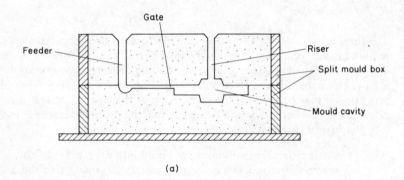

(a)

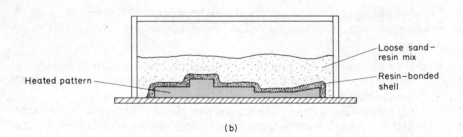

(b)

FIGURE 12.1   (a) *Mould for a simple sand-casting;* (b) *making a shell mould –*
loose sand–resin mix is dumped over a heated metal pattern

Another method frequently used for core making, and for moulds themselves, is the CO$_2$ process. Fine silica sand is used and this is thoroughly mixed in a sodium silicate solution before use. The sand, wetted with sodium silicate solution, is hard rammed to make the core, or mould, and then carbon dioxide is passed through the sand. The gas reacts with the sodium silicate, forming silica gel which effectively binds all the sand grains together. This gives a hard core with an erosion-resistant surface, enabling castings of good surface quality to be made. One disadvantage is that the CO_2- bonded sand, being very strong, is sometimes difficult to remove from solidified castings of intricate shape.

When the completed mould (with cores, if applicable) has been assembled, it is ready to receive the liquid metal. Liquid metal is carefully poured into the mould and allowed to solidify. When the metal has completely solidified the sand mould is broken up and the casting removed. 'Fettling', the operation to remove feeder heads, runners and riser heads, is then carried out, followed by any necessary machining operations and inspection.

Due to the low thermal conductivities of moulding sands, the rate of solidification within a sand mould is fairly low, and this results in a casting possessing a fairly coarse crystal grain structure. Most metals undergo a considerable volume shrinkage during solidification, and it is the function of the riser heads to provide reservoirs of liquid metal to feed this shrinkage. Adequate provision of risers should largely eliminate the possibility of major solidification shrinkage-zones within the casting, but finely divided inter-dendritic porosity is inevitable. Other defects that may occur in sand castings are sand inclusions, cold shuts, hot tears, and gas porosity. The major cause of sand inclusions within a casting is the washing away of loose sand from the walls of a poorly prepared mould. Cold shuts within a casting are a sign that the metal was poured at too low a temperature. Hot tearing, the fracture of a portion of the casting within the mould, is a result of tensile stresses being built up in parts of the casting due to thermal contraction, and is usually due to a poor design of the casting. The causes of gas porosity within the casting may be either pouring liquid metal with a high dissolved-gas content, or the generation of steam within the sand mould. This second type of porosity, known as reaction gas porosity, may occur either when the sand mould is too moist, or if the mould permeability is too low to allow any steam generated within the mould to escape to the atmosphere.

Despite its apparent disadvantages, sand casting is suitable for the production of castings in almost any metal and of almost any size from a few grams up to several hundred tonnes in mass.

Shell moulding is a technique for producing small moulds, or cores for sand moulds (see figure 12.1b). The mould, or core, pattern is made of metal and this is heated to about 250°C. Shell-moulding sand is 'dumped' over, or blown into, the heated pattern and the resin cures, binding the sand grains together. The mould shells produced in this way have a wall thickness of about 3 mm and are strong enough to be handled.

A sand mould, or shell mould, can only be used once, and while the production of small and medium-sized sand moulds can be largely automated, the process of sand casting does possess disadvantages. About 40 per cent of the total weight of casting produced is the weight of the feeder heads, runners, and risers, and this material has to be removed during fettling and returned for remelting. The surface finish obtainable is not particularly good and dimensional tolerances are not very high.

12.3 Diecasting

Some of the disadvantages associated with sand casting could be overcome by casting the metal into an accurately prepared metal mould, or die. With a metal mould the liquid metal may either be poured in under gravity alone, or it may be injected into the mould under pressure. Gravity diecasting is very similar in principle to sand casting, in that the die design must include riser heads to feed solidification shrinkage. The main advantages of the process are that a more rapid solidification rate produces a finer crystal grain structure than in a sand casting, and in consequence the diecasting will possess a higher strength than a corresponding sand casting. Also, better dimensional control and surface finish will be obtained, as compared with a sand-cast product. The cost of dies is high and the use of the process is restricted to metals and alloys with melting temperatures not exceeding values of about 1000 to 1100°C. Gravity casting into metal moulds is often referred to as permanent mould casting, and the term diecasting tends to be reserved for the pressure die process (see figure 12.2).

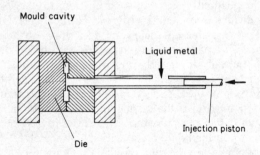

FIGURE 12.2 *Principle of cold chamber pressure diecasting machine*

The injection of liquid metal under pressure into a mould offers several advantages. It means that very thin sections can be cast, that metal can be forced into the recesses of a mould of very complex shape, and that rapid solidification under pressure will considerably reduce porosity. Also the need to provide risers in the mould is eliminated. In place of risers all that is necessary is a series of vents, of about 0.1 mm in thickness, at the parting line of both halves of the die. These will be sufficient to allow for the escape of air from the mould cavity as metal enters, but will be too small to permit the passage of metal. In diecasting there is little metal wastage and a casting is obtained that requires little, or no, machining. The process is, however, restricted to the lower melting-point metals and alloys, the principal ones being zinc, magnesium, aluminium, copper, and their alloys.

Figure 12.3 shows a number of precision die-cast parts in zinc and is a good illustration of how intelligent use of diecasting may save expensive machining-operations and permit easy assembly of components in a machine.

FIGURE 12.3 *A saw-grinding attachment making use of nine zinc-alloy castings; the figure shows the individual castings and the complete assembly – the knobs are die cast onto the adjusting screws. (Courtesy of Zinc Development Association and Hollands and Blair Ltd)*

12.4 Other Casting Processes

Another important casting process is the *investment*, or lost-wax, process. This process has its origins in the early Chinese and Egyptian civilisations, but it has become a valuable part of twentieth-century technology. Some metal components have to withstand very high temperatures and stresses in service, and in the case of such components as aero-engine turbine blades, the parts have to be made to very high standards of dimensional accuracy. Many materials sufficiently strong at high temperatures are virtually impossible to machine or shape at ordinary temperatures. A feasible solution would be to use a precision casting-process. Pressure diecasting is not possible in these cases, since the melting point of the alloy would be too high, and investment casting can provide the answer. In the investment process a master mould is produced in a readily machinable metal, such as brass, and this mould is used for the production of accurate patterns in wax. The wax patterns are then coated in a ceramic paste, or slurry, and when this has dried out and hardened, it is heated to melt out the wax and leave a perfect ceramic mould.

A method of casting that can be used for parts with an axis of rotational symmetry is *centrifugal* casting. This process is particularly suitable for the manufacture of large-diameter pipes for water and gas mains. A cylindrical mould shell, lined with moulding sand, is rapidly rotated about is longitudinal axis and liquid metal poured in. Under the action of centrifugal force, the liquid metal will spread evenly along the length of the mould. Rotational speeds are such as to give accelerations of the order of 60g. The properties of the finished casting are very good since the centrifuging action ensures that any slag particles, which are almost invariably less dense than the metal, segregate to the bore of the pipe, from where they may be easily removed by machining. Similarly, any gases will also escape to the bore leaving a sound casting. Centrifugal casting can also be used for producing cylindrical shell bearings in white metal, and in bronze and other copper alloys.

A comparison of the major casting processes is summarised in table 12.1.

TABLE 12.1

A comparison of casting methods

	Sand casting	Gravity diecasting.	Pressure diecasting	Centrifugal casting	Investment casting
Alloys that can be cast	Unlimited	Copper-base, aluminium-base and zinc-base alloys	Copper-base, aluminium-base and zinc-base alloys	Unlimited	Unlimited
Approximate maximum possible size of casting	Unlimited	50 kg	15 kg	Several tonnes	2 kg
Thinnest section normally possible (mm)	3	3	1	10	1
Relative mechanical properties	Fair	Good	Very good	Best	Good
Surface finish	Fair	Good	Very good	Fair	Very good
Possibility of casting a complex design	Good	Good	Very good	Poor	Very good
Relative cost for production of a small number off	Lowest	High	Highest	Medium	Low
Relative cost for large scale production	Medium	Low	Lowest	High	Highest
Relative ease of changing the design during production	Best	Poor	Poorest	Good	Good

12.5 Ingot Casting

In very many instances it is necessary to shape metals by deforming them plastically. The starting point for these working processes is a large regular-shaped casting, called an *ingot*. Ingots for further working are generally square, rectangular, or circular in cross-section, and may vary in size from about 10 kg for some non-ferrous alloys, up to 100 t for some large steel forging ingots.

Ingots may be cast by pouring the liquid metal into large permanent moulds made from cast iron. Cast-iron moulds for the production of steel ingots are tapered to allow for easy removal of the solidified ingot from the mould, and are termed either 'big-end up' or 'big-end down' depending on the direction of the taper. At the end of the steel refining stage the liquid steel is in a highly oxidised state. Additions are made to the bath to deoxidise the steel before pouring, or teeming, the steel. Fully deoxidised steel is termed 'killed' steel, and most killed steels are poured into moulds of the big-end-up variety. In order to minimise the formation of pipe defect due to solidification shrinkage, the moulds are generally provided with a refractory collar, or 'hot top' to act as a reservoir for liquid metal to feed the shrinkage (see figure 12.4). Liquid metal poured into a large mould from the top can splash considerably, and the splashes would freeze instantly on the upper mould walls. The splash surface would then rapidly oxidise and not weld satisfactorily to the main body of the ingot when the mould is filled with liquid metal. This would give rise to surface defects in the subsequent rolled or forged product. In order to avoid this, some steel ingots are bottom poured. This process, which is considerably more expensive than straight teeming into moulds, is normally only used for the higher-grade steels.

Some grades of low-carbon, or mild steels, are not deoxidised before teeming. These are called 'rimming' steels and they are normally cast into moulds of the

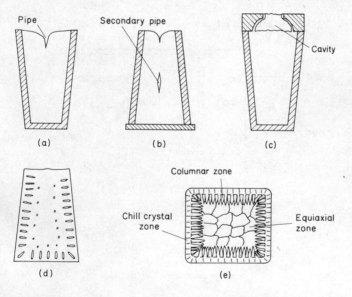

FIGURE 12.4 *Ingot casting:* (a) *big-end-up type;* (b) *big-end-down type, showing secondary pipe;* (c) *big-end-up with hot collar;* (d) *section of rimmed steel ingot;* (e) *transverse section of ingot*

big-end-down type. There is considerable gas evolution during the solidification of rimming steels. When solidification first begins at the mould walls, carbon monoxide is evolved according to the reaction

$$FeO + C = CO + Fe$$

The stream of gas bubbles tends to carry impurities away with it and the outer layers of the ingot solidify as almost pure, clean iron, with the impurities concentrated in the central portion of the ingot. Some of the gas bubbles are trapped between growing crystals of iron giving a widely distributed porosity, and this porosity compensates for the solidification shrinkage of the metal. If the internal surfaces of the gas pores are clean they will readily weld together under rolling pressures, giving a sound product. Rimmed steel is very suitable for the production of mild-steel sheet because the pure-iron rim gives a ductile skin of good surface quality, and this makes the steel ideal for deep drawing and pressing operations.

Large ingots in permanent moulds possess coarse grain structures, and there will also be variation in crystal structure across the ingot. A typical transverse section of an ingot is illustrated in figure 12.4e. There are three main crystal zones shown. The initial chill zone, which is of small thickness, possesses small crystals. The second zone is the columnar zone, and this contains large crystals that have grown preferentially inwards from the mould wall along the direction of the major temperature gradients. Finally, when the temperature of the central liquid zone has fallen to around the freezing point of the metal and temperature gradients are small, randomly orientated or equiaxial crystal grains form.

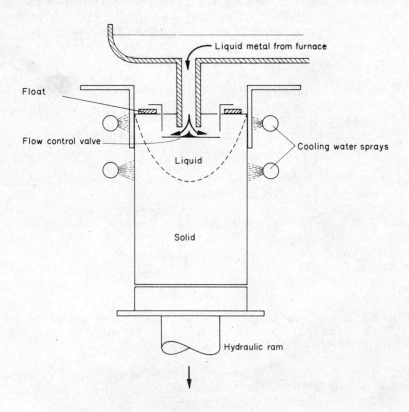

FIGURE 12.5 *Semicontinuous casting of aluminium*

Before 1940 ingots of all metals were cast, almost exclusively, into permanent cast-iron moulds, but considerable attention was given to the development of casting processes that would give ingots of a better and more consistent quality, and in particular, a casting process that would be continuous in operation. The introduction of continuous casting, on a production scale, first took place in the aluminium and copper-alloy sectors of industry. The continuous casting of steel was a much later development, due to the greater difficulties that had to be overcome, with molten-steel temperatures of about 1550°C, compared with temperatures of about 700°C and 1000°C respectively for molten aluminium and molten brass. Today, almost all large ingots produced in the aluminium industry, and much of the ingot production in the copper industry, are continuously cast. A growing proportion of steel-ingot production is also cast in this way.

Figure 12.5 shows the lay-out of a semicontinuous casting unit for the production of aluminium ingots. The mould is shallow, and the distance from the metal surface to the bottom of the mould skirt is about 75 mm. The ingot withdrawal rate must be carefully controlled, and this is done by means of a hydraulic ram. Withdrawal speeds vary considerably with alloy composition and the sectional size of the ingot and may be between 20 and 200 mm per minute. The liquid metal solidifies very quickly and this means that not only does the ingot possess a fairly fine grain size, but also that segregation of impurities is largely eliminated and there is no possibility of piping defect occurring. In semicontinuous casting the maximum length of ingot possible is determined by the length of stroke of the hydraulic ram. This is usually of the order of 5 m. The fully continuous casting process is similar, but in this case steady withdrawal of the ingot from the mould is accomplished by having a pair of pinch rolls situated about 1 m below the mould. The continuous ingot is cut into suitable lengths by a high-speed saw.

12.6 Hot Forging and Rolling

As stated in earlier chapters, metals generally show elastic behaviour up to a certain level of deforming force but they may be deformed in a plastic manner when stressed beyond the elastic limit. The ability of metals to be deformed plastically is made use of in many manufacturing processes. These plastic-deformation processes for the shaping of metals are termed working or wrought processes. The working of metals may be undertaken at ordinary temperatures or the workpiece may be heated. As the temperature of a metal is increased so the value of the elastic limit for the material is decreased. This means that a smaller deforming force is required to shape a hot piece of a metal than that necessary to shape the same metal when cold.

When a metal is deformed plastically work hardening occurs and it becomes progressively more difficult to continue the plastic deformation, but if the material is heated above its recrystallisation temperature (see section 8.5) deformed metal crystals will reform in a strain-free manner. This means that if a metal is worked at a temperature above its recrystallisation temperature, not only is a lower deforming force necessary because of the reduced elastic limit value, but continued plastic deformation may be performed because strained crystal grains are continually recrystallising.

The term 'hot working' is used to describe those shaping processes that are carried out at temperatures above the recrystallisation temperature of the metal and the term 'cold working' describes processes in which the metal becomes work hardened to some extent. It should be noted that if mild steel is worked at a temperature of 300 or 400°C this is technically cold work, since the recrystallisation temperature of cold worked mild steel is about 500°C, whereas pure lead, with a recrystallisation temperature of about 0°C may be hot worked at ordinary temperatures.

Almost without exception, cast metal ingots are hot worked as the first stage or stages of shaping, and for some products the finished shape is produced by hot work. The three major processes for the hot break-down of cast ingots are forging, rolling, and extrusion. The ingots are preheated in soaking pits or furnaces to bring them up to the desired temperature, usually about 100–200°C below the melting range. At such a high temperature the metal is much more plastic and requires smaller deformation forces than would be necessary at low temperatures. During the first stages of hot deformation the coarse 'as cast' structure of the ingot will be broken up and distorted, but almost immediately recrystallisation will occur and further deformation can be carried out. A sequence of rolling passes or forging blows will generally be planned so that the final deformation takes place at a temperature close to the recrystallisation temperature (the critical temperature range, in the case of steels) so that recrystallisation without excessive grain growth occurs.

— Forging means the shaping of metal by a series of hammer blows and it is a process that may be used for the shaping of both large and small components. The simplest example is a blacksmith's forging of a hot piece of metal by hammering the workpiece on an anvil. Heavy smith's forging is fundamentally similar, differing only in the scale of the operation. The workpiece may be an ingot of 100 tonnes weight and the deforming force provided by a massive forging hammer, but the whole process is controlled by the master smith, who decides each time where, and with what force, the blow should take place. Another form of forging is closed-die forging, in which the hammer and anvil possess shaped recesses that effectively form a complete mould. The hot metal workpiece is caused to flow, under pressure, into the die cavity to produce the desired shape. This technique is used for the mass production of small forgings, for example, connecting rods for internal-combustion engines.

The way in which the metal flows during a forging operation is of considerable importance. The forged metal will possess a flow pattern or *fibre structure*. This is briefly discussed in section 12.8 and it will be seen that the flow pattern within a forging can have a major influence on the properties of the component.

In hot-rolling operations the ingot is passed between two large cylindrical rollers, and the roll surfaces have to be provided with flood cooling to prevent overheating of the rolls and possible welding between ingot and rolls. Water, or a water/soluble-oil emulsion, is used, depending on the nature of the material being rolled. If the rolls are plain cylinders the ingot can be processed into flat plate material. Rolls with grooves machined into the surface are used for the production of round bars and sectional shapes, such as rails and structural beam sections.

The product of the hot-rolling mill may be the finished product, requiring only trimming to size, or it may be an intermediate product that will be deformed further by cold-working operations. Structural plate and sections are examples of products used in the hot-worked condition.

A major advantage of any hot-working process is that it breaks up the coarse grain structure of a cast ingot. Also, under the pressures involved, clean cavities such as minor shrinkage porosity will be closed up. Consequently the strength of a hot-worked metal product will be considerably better than that of a casting. Some disadvantages of hot working are that very close control of dimensions is not always possible, and the product will have a poor surface finish due to the effects of high-temperature oxidation. Subsequent cold-working operations will give close dimensional tolerances and a good, clean surface.

12.7 Extrusion

The process of extrusion is used for the shaping of some metals, thermoplastic materials and also plastic clays. The principle of extrusion is to force material in a plastic state through a shaped orifice or die in a similar manner to the flow of

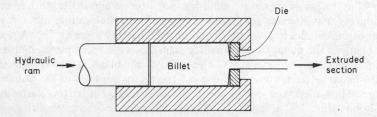

FIGURE 12.6 *Principle of extrusion*

FIGURE 12.7 *Typical examples of extruded sections. (Courtesy of the British Aluminium Co. Ltd)*

toothpaste from a tube. There are, however, certain differences between the
extrusion of metals and plastics materials. In the hot extrusion of metals an ingot
of circular cross-section (in industry such an ingot is termed a *billet*) and preheated
to about 100–200°C below its melting temperature, is inserted into the chamber
of a hydraulic extrusion press. The hot ingot is then forced to flow through the
shaped die. The principle of direct extrusion is shown in figure 12.6. It is possible
to extrude an almost infinite variety of sectional shapes, including complex hollow
sections, from solid ingots. Some indication of the versatility of the process is given
in figure 12.7.

The principal metals that are hot extruded are aluminium and its alloys
(extrusion temperatures 400–550°C) and copper alloys (extrusion temperatures
650–1000°C). The hot extrusion of steel poses special problems due to the very
high temperatures necessary. At the temperatures involved, namely 1100–1250°C,
die wear could be excessive. Glass is used as a lubricant for steel extrusion. This is
applied by placing a glass-fibre mat against the die and covering the hot billet with
a glass-fibre sleeve before placing it in the machine chamber.

Impact extrusion of metals is a cold-forming process and is the method used
for producing such items as toothpaste tubes and other collapsible tubes in lead or
aluminium, cigar tubes in aluminium and dry-cell battery cases in zinc. A small
slug or blank of the metal to be formed is placed in a die and a punch delivers a
very high impact force to the blank. The metal of the blank is forced through the
small clearance between punch and die and travels up the sides of the punch
forming a thin-walled tube (see figure 12.8). Zinc blanks for battery cases are
normally heated to 150°C before impact extrusion since this hexagonal-structured
metal has little ductility at ordinary temperatures.

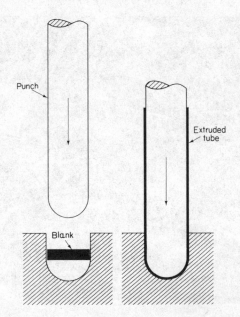

FIGURE 12.8 *Impact extrusion of a cigar tube*

Thermoplastic materials may be extruded with ease and the process, unlike that for metals, is fully continuous. The material, in a granulated or powder form, is fed from a hopper into the cylinder of an extrusion machine. A screw feed mechanism is used to carry the material along the cylinder through a heating zone to the die. During its progress through the cylinder the material becomes compressed and fused into a homogeneous plastic mass. The temperature of extrusion is in the range 100–180°C, depending on the nature of the material. The principle of a plastics-extrusion machine is shown in figure 12.12b.

12.8 Fibre Structure

The final properties of a hot worked metal product depend to a certain extent on the type of deformation process that was used. Commercial materials are not completely pure and will contain, in addition to the alloying and trace elements, small non-metallic particles derived from the fluxes and slags involved in the melting operations. Within the cast ingot these particles have a fairly random distribution. When the ingot is deformed these foreign particles will flow with the metal and will form fibre lines within the structure. After each hot deformation the flowed metal crystals will recrystallise, but the fibre particles will remain

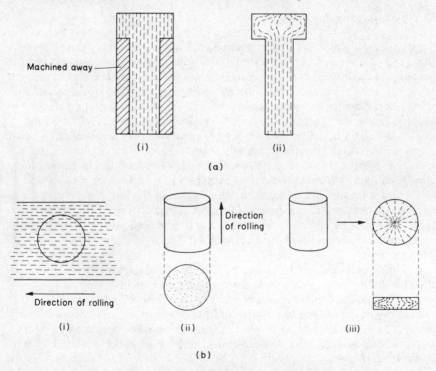

FIGURE 12.9   *Fibre lines:* (a) *bolt shaped by* (i) *machining from rolled stock,* (ii) *forged bolt head;* (b) *gear blank formed by* (i) *blanking from rolled plate,* (ii) *parting off from rolled bar,* (iii) *forging from small-diameter bar*

unchanged. In rolled and extruded products all fibre lines will lie parallel and will be in the direction of working, but in a forging the metal-flow pattern will be more complex. Figure 12.9a shows the fibre structure of two bolts, one of which has been formed by machining from a solid hot-rolled bar, and the other which has been formed by forging the bolt head from small-diameter hot-rolled bar stock. The difference in fibre structure will give the components differing properties even though all other factors may be equal. Another example of different production routes giving different fibre structures is the case of the small gear blank (see figure 12.9b) formed by (i) blanking from hot-rolled plate, (ii) parting off a thin slice from a large-diameter hot-rolled bar, and (iii) parting off a length from a small-diameter hot-rolled bar and forging this into a thin disc. The fibre lines may constitute lines of weakness within the material, and lines along which fracture may occur. In the case of the gear blank formed from rolled plate the fibre lines are tangential at two points on the disc and teeth in these positions could be subject to failure. In case (iii) all fibre lines within the disc are radial and the risk of failure of the gear is reduced. The term 'upsetting' is given to the type of forging that would be used in case (iii), since the former flow pattern is completely reversed and the arrangement of the fibre lines is totally altered.

12.9 Cold-forming Processes

Cold working is plastic deformation performed at temperatures below the recrystallisation temperature of the metal. During cold working the crystal structure becomes broken up and distorted and the material becomes strain, or work, hardened. The mechanical strength is increased by cold working, and the material becomes harder, but more brittle. The electrical resistivity is also increased. Eventually, the metal becomes so hard that further attempts at cold working would cause fracture. After cold working the material may be softened by heating it to a temperature above the recrystallisation temperature and allowing the distorted crystal grains to recrystallise. This type of process is termed *annealing*. Very good dimensional control can be exercised in cold-working processes and good surface finishes can be achieved. The surface of the hot-worked product may be covered by an oxide scale that was formed at high temperature. This layer of oxide scale must be removed before cold-working operations can be commenced. Oxide scale may be removed from hot-worked steel by pickling the steel in hot sulphuric acid.

Cold rolling is used for the production of sheet and strip material. The majority of cold-rolling operations are these days carried out on 4-high strip mills (see figure 12.10). (The term strip is applied to any material rolled from coils, irrespective of the width.) In a 4-high mill the two small-diameter work rolls are power driven. The large backing rolls are for transmitting pressure to the work rolls and for preventing excessive deflection under load of the work rolls. After cold rolling, the coiled strip may either be levelled and cut into sheet, or circular blanks, of the desired size, or supplied as coil for some other manufacturing process. It is possible to cold roll some materials down to extremely thin foil gauges. The aluminium foil that is used for chocolate and cigarette wrappings is 0.008 or 0.009 mm in thickness and it has been produced by a cold-rolling process.

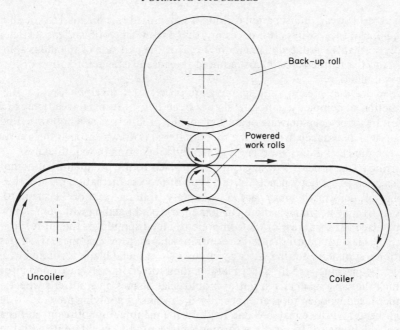

FIGURE 12.10 *4-high strip mill*

The final properties of cold-rolled material may vary between those of the fully work-hardened condition and those of the annealed, or soft condition. Properties intermediate between these two extremes may be obtained by carefully controlling the amount of cold reduction given to the material after an annealing operation (refer to table 8.1).

Another useful cold-working process is drawing. This is used for the production of wire, rod, and tubing. The annealed material is pulled through a die with a reducing diameter (see figure 12.11a). The die materials must be very hard to

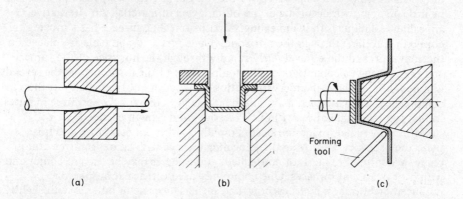

FIGURE 12.11 *Cold-forming processes:* (a) *wire drawing,* (b) *deep drawing,* (c) *spinning*

resist abrasive wear, and tungsten carbide is often used as a die insert. Not all metals can be drawn successfully. It is necessary that the metal work hardens sufficiently rapidly so that the reduced-diameter material at the exit side of the die is able to withstand the tensile force that is required to cause plastic deformation of the undrawn metal.

Deep drawing, pressing, and spinning are processes that can be used for the production of complex and hollow shapes from flat sheet metal (see figures 12.11b and c). In a true deep-drawing operation for the production of a hollow cylindrical shape, such as a saucepan body, the sheet material is drawn through the aperture between punch and die, and there is generally a reduction in wall thickness occurring in the process. Pressing is the term given to the production of complex shapes, such as car-body panels. In this type of process the nature of, and the amount of, deformation may vary considerably from one part of the shape to another. It may be simple bending in some parts, and bending with some drawing taking place in other areas of the component. Metal spinning is the most suitable means of making conical shapes or re-entrant shapes, providing there is an axis of rotational symmetry for the latter. A circular sheet-metal blank is rotated in a lathe-type machine, and the sheet metal is then forced into shape around a former, in much the same manner as a potter moulds clay shapes on a potter's wheel. Leather-faced wooden tools are normally the means of applying pressure to the metal sheet, and these may be hand held. In the manufacture of re-entrant shapes, such as the bodies of teapots, the former is segmented to facilitate its withdrawal from the moulded shape.

In many instances it is necessary for a cold-worked metal to be annealed, either as an intermediate stage in processing, or as a final process. In most cases the annealed material is required to have a fine crystal grain size, so that recrystallisation without excessive grain growth is necessary. The two principal types of annealing process are *batch* annealing and *flash* annealing. In the former process a large batch of material is placed in the furnace and annealed in one lot. For this type of annealing the maximum temperature in the cycle is only marginally above the recrystallisation temperature. The surface layers of a batch furnace load may reach the annealing temperature several hours before the inner layers. With only a small temperature excess above the recrystallisation temperature, there will be little grain growth occurring in the outer layers of material. An alternative annealing technique is flash annealing. A flash-annealing furnace has a moving conveyor travelling through the hot zone. Single metal components are placed on the conveyor, and these travel fairly rapidly through the hot zone of the furnace and then through a cooling zone. The temperature of the hot zone may be several hundred degrees above the recrystallisation temperature of the metal, but the components being annealed may only be in the hot zone for two or three minutes, so that full recrystallisation without excessive grain growth occurs.

With many metals oxidation occurs rapidly at high temperatures, and it is therefore necessary to provide the annealing furnace with an oxygen-free atmosphere. Many types of inert, or controlled, atmospheres may be used in conjunction with heat-treatment furnaces. One commonly used furnace atmosphere is produced by burning a hydrocarbon fuel, such as propane or butane, with slightly less than the theoretical quantity of air required for complete combustion. The burnt gases are then passed into the furnace and oxygen excluded.

12.10 Moulding of Plastics

Polymerisation reactions can take place by merely allowing a monomer to stand at the reaction temperature, with or without a catalyst present. If polymerisation is allowed to take place in this manner, known as mass polymerisation, the mass becomes increasingly viscous as the reaction proceeds. Mass polymerisation can be used effectively for the production of some materials, but it is not suitable for all thermoplastic polymer production. Polymerisation reactions are exothermic, and as the reaction mass becomes more viscous it becomes very difficult to stir and to dissipate heat. Consequently it becomes difficult to control the average molecular weight of the polymer. Many thermoplastics are produced by polymerisation in solution, or in water emulsion.

The polymer formed is then ground into powder or broken up into granules, but in order to produce a suitable moulding material it is then usually compounded with plasticisers, lubricants, and pigments. It is necessary to add a plasticising agent to many polymers to improve both their moulding performance and toughness. Plasticisers are generally non-volatile solvents that partially dissolve the polymer and increase the flexibility of the material. Another form of plasticising agent is a non-solvent oil, highly dispersed throughout the material.

Some thermoplastics may be cast into shape in moulds and allowed to solidify, but this process is not widely used. It is, however, suitable when only a limited number of items of any particular design is required and the cost of manufacturing an expensive die for injection moulding is not justified. Sheet material may be made by a process called casting, but this is not casting in the true sense. It involves pouring a solution of the plastic material onto a moving conveyor; the solvent subsequently evaporates, leaving behind a solid sheet.

Injection moulding is a very widely used process for the production of shaped components in thermoplastic materials, and it is a process analogous to the diecasting of metals. The compounded plastic material is injected into a die chamber, where it is allowed to cool, and it is then ejected when rigid. Figure 12.12a shows a typical ram injection-moulding machine. The plastic is conveyed through a heating zone − the plasticising zone − by the ram mechanism. The heating and compression combine to produce a homogeneous plastic mass. Many injection-moulding machines have helical-screw feed mechanisms to move the plastic material through the heating zone. The shearing effect on the plastic due to rotation of the screw aids rapid plasticising of the material. In this type of machine the whole screw is also capable of moving longitudinally within the barrel of the machine in order to ram each injection shot into the mould.

Rod, tubing and film can be produced by extrusion. In this process the raw plastic is fed into the cylinder of the machine and propelled through the heated zone by a screw feed mechanism and forced out through the extrusion die (see figure 12.12b). It is also possible to extrude directly around wire or rod to produce insulated wire.

Developments of extrusion are blow moulding and film blowing. Bottles and other hollow shapes are produced by blow moulding. Tubing is extruded into a shaped mould and is then inflated by air to fill the mould cavity. Thin film is made by a film-blowing technique, the principle of which is shown in figure 12.12c. Another process that can be used for the production of film and sheet in thermoplastic materials is calendering. In this process a pre-mixed heated plastic dough is passed through a series of rollers (figure 12.12d). Plasticised PVC sheeting, in thicknesses ranging from 0.1 to 0.7 mm, is manufactured in this way.

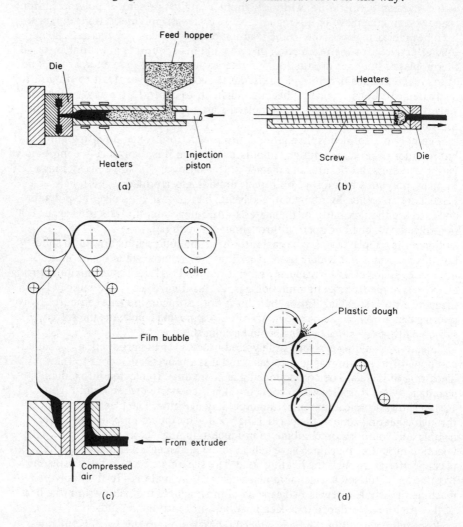

FIGURE 12.12 *Forming of thermoplastics:* (a) *injection moulding;* (b) *screw extruder;* (c) *principle of film blowing;* (d) *principle of calendering*

Thermoplastic sheet material may be formed into a variety of shapes by several means. One of the most common methods is vacuum forming. In this system heated flat sheet is clamped over a mould box and the intervening space is evacuated. This method can be used for forming comparatively large shapes such as baths and car-body panels as well as for small shapes in very thin sheet, such as thin compartmented trays for use in packaging.

The thermosetting plastics are produced by condensation polymerisation and the reaction must be stopped before completion, since the raw plastic for moulding is required in a part-polymerised condition. The polymerisation is completed during the moulding process to give a full network rigid structure. This hardening within the mould is termed *curing*. Curing is accompanied by a volume contraction of the resin and an inert filler must be mixed in with the resin powder in order to reduce the effect of this shrinkage. The use of a filler will also tend to reduce the cost of the moulding material and many fillers also give an improvement in some properties. Other additives used in the compounding of thermosetting moulding powders are catalysts to accelerate the curing process, pigments to colour the resin, and lubricants to prevent the material from sticking to the moulds. A variety of materials may be used as fillers and the filler generally accounts for between 50 and 80 per cent of the total weight of moulding powder. Coconut-shell flour and wood flour, a fine soft-wood sawdust, are two widely used fillers. Wood flour increases the impact strength of the material. Coconut-shell flour, a very cheap filler, would, if used alone, give rise to mouldings with comparatively poor properties, and so it is generally used in conjunction with wood flour.

Cotton flock and chopped fabric are often added as fillers where a high resistance to impact is required. Paper pulp and nylon fibre may also be used for this purpose. Other fillers that may be used to improve specific properties are asbestos, for improved resistance to heat, mica, for improved electrical resistance, and graphite, to reduce the coefficient of friction and give good sliding properties.

The principal forming processes for thermosetting plastics are compression moulding and transfer moulding. In compression moulding, the moulding powder is compressed between the two parts of a heated metal mould, or die. The powder becomes plastic and flows into the recesses of the mould. The pressure is maintained until the polymer has cured, that is, until full polymerisation has occurred, and the material has set rigid. Compression moulding is not suitable for the moulding of shapes of thick section, or with large changes of section. The resins possess low thermal conductivities, so that the centre portion of a thick section may not become fully heated and only partially cure. This problem is largely overcome in transfer moulding. The resin powder is heated in an antechamber and when plastic it is injected into the main mould, where curing occurs (see figure 12.13). It is, however, a more expensive process than simple compression-moulding since there is some scrap material formed each time.

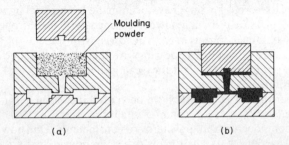

FIGURE 12.13 *Transfer moulding:* (a) *moulding powder in antechamber;* (b) *resin forced into mould for curing*

The cold-setting rigid plastics may be readily formed into shapes by casting. These castable materials may be used for encapsulating small electrical components, and for mounting metallic and biological specimens.

The cold-setting polyester and epoxide resins, which are frequently reinforced with glass fibre to give strong and resilient materials ('fibreglass'), are formed to shape, in most instances, by hand lay-up techniques. In this process, a mould, or former, is prepared and given a non-stick coating. A glass-fibre mat is then laid, by hand, over the former and the liquid resin applied by brush or spray. Further layers of fibre mat and resin may be added to build up the thickness. The component is then allowed to cure and harden. Curing time may be accelerated by heating the lay-up slightly.

Expanded, or foamed, plastics and rubbers are used to a considerable extent for packaging, heat and sound insulation, and as cushioning materials. Foams may be formed by a number of methods, including fully aerating an uncured resin in liquid form, or mixing the plastic resin with a volatile material that vapourises at the temperature of the curing process. Another method of producing a foam is to add a chemical blowing-agent. This is a chemical that decomposes with the evolution of a gas, such as carbon dioxide, when it is heated; the technique is used for the production of expanded polystyrene.

12.11 Powder Techniques

Components of simple or intricate shape may be produced by the compaction of fine powder in a die, followed by sintering the compact at an elevated temperature to give a strong coherent article. Fabrication from powder is one of the few processes available for the production of ceramic parts, and it is also a very useful method for the production of small metal components.

Manufacture from powder is a very useful process for the production of parts in metals of very high melting point. It is both difficult and expensive to melt these metals. Tungsten, for example, melts at $3410°C$, but a powdered tungsten compact may be sintered into a fully coherent part at temperatures of the order of $1600°C$. Another useful application of powder metallurgy is in the production of 'hard metals', namely sintered mixtures of hard metallic carbides and powdered cobalt, nickel, or other strong metals. These hard metals are used for the manufacture of cutting-tool tips, percussion tools, and die inserts.

For many years porous bearings have been made from metal powders. If a fairly low compaction pressure is used, a powder compact with a controlled degree of porosity can be obtained. Porous bronze bearings, with some 30 or 40 per cent porosity are made in this way. The bearings are then soaked in lubricating oil before they are put into service, and they can give long periods of service without requiring attention.

In recent years powder processes have been used increasingly for the production of small machine components of complex shape (see figure 12.14). The advantages of fabricating components from powders are considerable. One powder-compaction operation produces an accurately dimensioned component that does not require machining, and there is no scrap material produced. The only other operation necessary to produce the completed component is sintering. In order to produce a similar component by conventional methods it may be necessary to start with stock material produced by hot-working and cold-working processes and to give this stock several expensive machining operations with, in consequence, some scrap arising at each production stage. Although the cost of metal powders is high in comparison with the cost of ingot metal, it may well be cheaper to produce components by powder-metallurgy techniques than by conventional machining.

Conventional processes involving melting and casting, followed by hot and cold working and machining involve the use of a very large amount of energy per unit mass of finished production. The days of cheap energy supplies that existed prior to 1974 are most unlikely to return. This, together with the need to exercise some restraint over the exploitation of fossil-fuel reserves, should mean that low-energy-consuming processes such as powder-metallurgy techniques will assume very great importance in the future.

FIGURE 12.14   *Typical examples of parts made by powder metallurgy.*
(Courtesy of Sintered Products Ltd)

There are several methods available for the production of metal powders, but not all methods apply to all metals. The more brittle metals may be powdered by pulverising in a hammer mill. Softer metals may be converted to powder by atomising a stream of the liquid metal in a jet of air or an inert gas. Many metals, including aluminium, iron, tin and zinc, are converted to powder in this way. Magnesium and zinc, metals with comparatively low boiling points, are sometimes obtained in powder form by condensing the metal from the vapour phase. Some pure metal powders are obtained by electrolytic deposition of metal from a solution of a salt of the metal. The electrolytic-cell operating conditions are arranged so that the cathode deposit is soft and spongy, rather than a hard coherent plating. This method is particularly suitable for the production of copper powder. Chemical methods for the production of metal powders include the reduction of heated powdered metallic oxide to metal in a stream of a reducing gas, such as hydrogen, and the dissociation of volatile compounds, such as carbonyls. (Some metals combine with carbon monoxide to form gaseous compounds, called carbonyls.) Heating the carbonyl causes it to dissociate into metal and carbon monoxide. The carbon monoxide may be recirculated. Pure nickel and iron powders can be made in this way.

The metal powders are compacted in alloy-steel, or cemented-carbide, dies at pressures of up to 750 MN/m^2. The pressure used depends on the nature of the metal powder, the average particle size, and the degree of porosity required in the compact. During the compaction operation there will be some plastic deformation of individual powder particles taking place, allowing for some mechnical interlocking of particles. There will also be some cold-pressure welding of particles occurring, and the compact will therefore possess sufficient strength to be handled. The compacts are then sintered in an inert atmosphere at some temperature below the melting point of the metal or alloy. The sintering temperature for iron is about $1100°C$, and for bronze it is about $800°C$.

Newer developments in powder technology include the production of sheet and strip material by compacting metal powders between rollers. This has been successfully accomplished, on a pilot-plant scale, for pure aluminium, some aluminium alloys and for mild steel. This technique probably has great potential since it offers a short route from raw material to finished product with a considerably lower level of capital investment than in conventional metal-processing plant.

Many of the newer oxide and carbide industrial ceramics are shaped from powder by a process of compaction and sintering, in the same general way that powder-metal parts are prepared. One major difference is that the ceramic powders are non-ductile and show no plastic deformation when compressed. The ceramic powder is often mixed with a small amount of a binder and a lubricant before moulding so that a coherent compact, which may be handled, is obtained. The ceramic compacts produced almost invariably contain some porosity. This porosity may be 'open', with interconnected pores, or it may be of the 'closed' type, with a series of isolated small pores. The pores act as flaws and stress raisers, and points for the initiation of fracture. Considerable shrinkage occurs during the sintering, or firing, operation, and this may be as high as a linear shrinkage of 20 per cent for some materials. Consequently sintered ceramic parts can rarely be produced to close dimensional tolerances. Under the most favourable circumstances the fired dimensional tolerances would not be better than ± 1 per cent for most materials. It is possible to shape ceramics accurately after they have been fired by a series of grinding operations, but this is both difficult and expensive owing to the very hard and brittle nature of these materials.

13

Cutting Processes

13.1 Introduction

Cutting and other machining processes are used to a very considerable extent for the shaping of materials. One of the biggest advantages of machining is the extreme flexibility of the processes. Shapes of any description can be obtained by intelligent use of the various machining methods available. In addition, extremely high dimensional-accuracy can be achieved. It is also possible to produce components with a variety of surface finishes. On the other hand, machining operations are high-cost processes. Considerable skilled labour and machine time are required per component manufactured, although for large-scale production, costs can be reduced by using multi-purpose numerically controlled machine-tools. Another economic disadvantage is that the material removed during machining operations, the machine-tool scrap, or swarf, is bulky and of comparatively low value. Compressed bales of swarf can be recycled but melting losses tend to be high. For economy in the production of components in large quantity it is best if the desired shape is almost completely obtained by casting or some plastic-deformation process and machining operations are reserved for finishing, so necessitating removal of the minimum amount of material.

13.2 Action of a Single-point Cutting-tool

Machining is really a cold-working process in which a cutting tool forms chips, due to a series of fractures at the surface of the material being cut. The ease with which a material may be cut and the type of surface finish obtained depend on many factors. These include the cutting-tool design, the depth and speed of cut, the nature of the material being cut, the type of cutting-tool material and the method of lubrication. Consider a single-point cutting-tool. The point of the tool is designed to give a *rake* angle and a *clearance* angle (see figure 13.1a). During a cutting stroke the wedge-shaped tool will exert a force F on the material. A shear plane is established along the line AB. The material just in front of the cutting tool will become very highly stressed and is subjected to plastic deformation (see figure 13.1b). If the material is soft and ductile there will be very considerable plastic deformation and a continuous chip will be formed. As the tool moves through the material and the chip moves upwards against the tool face there will be considerable heat generated due to both friction between chip and tool and work done in plastically deforming the material.

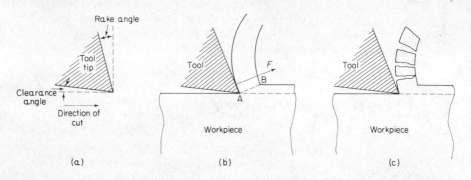

(a)   (b)   (c)

FIGURE 13.1   *Single-point tool:* (a) *rake and clearance angles,* (b) *cutting action,* (c) *discontinuous chip formation*

The rake angle of the tool is of great importance. Figure 13.2 shows sketches of two cutting tools, one with a small rake and one with a large rake, both taking the same depth of cut. It will be seen that the size of the shear plane AB is considerably greater for the small-rake-angle tool than for the tool with a large rake-angle. The magnitude of the force necessary to drive the tool through the material will be greater for the small-rake-angle tool than for the other. Because of the increased force necessary, there will be increased friction between chip and tool face for a small-rake-angled tool. The greater frictional heating will lead to increased tool-wear. Chip formation will tend to be discontinuous since the very severe work-hardening will lead to chip fracture. Theoretically a tool with a very large rake-angle would be suitable for cutting a soft and ductile metal. The action of such a tool would be similar to that of a hand chisel on wood. In practice a compromise has to be achieved between rake angle and the need to have a large enough wedge angle to give the tool sufficient strength.

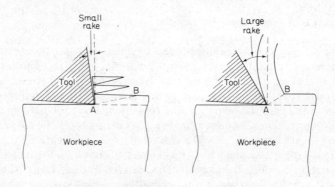

FIGURE 13.2   *Effect of rake angle*

If a material of low ductility is machined the low-ductility material will tend to fracture rather than plastically deform to a large extent and a whole series of small discontinuous chips will be formed.

Cutting fluids are frequently used in machining processes and they perform several functions. In addition to acting as a coolant for the tool and workpiece they reduce friction and help to carry away chips from the cutting point. This gives increased tool-life and leads to an improved surface-finish of the workpiece. Soluble oil emulsions are widely used as cutting fluids.

13.3 Machinability

As stated above a highly ductile material will give rise to a long continuous swarf chip when cut. The chip is deformed into a curve by the action of the cutting tool and so the continuous chip tends to form into a coil. Continuous swarf chips that may coil around the tool or workpiece can be very troublesome, especially in high-speed machining operations. With a less ductile material, rapid work-hardening leads to chip fracture and the formation of small discontinous chips. This latter type of chip formation is desirable for high-speed machining with automatic machine-tools. In general a decrease in the ductility of a metal is accompanied by an increase in hardness. For ease of machining, a material should not be too hard or the cutting material will not be able to penetrate the workpiece.

Some very soft and highly ductile materials spread under tool cutting-pressures. In these conditions the cutting tool tends to become buried in the workpiece and a tearing, rather than a clean cutting action, occurs. The machinability of a metal can be improved by any means that will decrease its ductility and susceptibility to fracture. This can be achieved by work hardening or alloying to give a harder alloy.

Multiphase metallic alloys generally possess good machinability. The presence of two or more phases within a metal structure means that there are discontinuities within the material and these assist in the formation of discontinuous swarf chips. Additions of insoluble phases are frequently made in the production of alloys to make free machining varieties of otherwise ductile materials. A good example of this technique is the addition of a small percentage of lead to brasses and mild steel. The presence of small globules of lead in the microstructure of these alloys provides small structural discontinuities without greatly affecting the ductility. Free machining mild steels may also be made by increasing the sulphur content. In this case, globules of manganese sulphide appear in the structure. Grey cast-irons possess good machinability. This is due to the presence of graphite flakes within the structure of the iron. The graphite also functions to some extent as a lubricant.

For the best machining characteristics the requirements are low hardness coupled with low ductility. Magnesium and its alloys provide a very good example of this combination of properties and these materials possess excellent machinability.

Many thermoplastic materials can be machined. Because of their very low thermal conductivities there could be overheating of the materials leading to some melting and charring and a poor surface finish. Water or soluble-oil cooling is necessary to prevent this. Thermoplastics tend to give continuous swarf and they are best cut by tools with zero rake or negative rake. Thermosetting materials are not generally machined with the exception of phenol-formaldehyde laminates ('Tufnol'). The machining characteristics of these materials are roughly comparable to those of brasses.

Ceramic materials are of high hardness and no ductility and consequently it is impossible to shape them by cutting. The only exception to this is the intermediate stage in the production of silicon nitride. After the compaction of silicon powder the compact is in a suitable form to be machined by conventional cutting-processes such as turning and milling. After shaping the compact is reaction-sintered in nitrogen. The very hard silicon-nitride ceramic is formed with no dimensional changes. Apart from this exception ceramics may only be processed by grinding.

13.4 Cutting-tool Materials

One of the prime requirements of a cutting-tool material is hardness. Most very hard materials are also brittle, but it is necessary that a cutting tool possess considerable toughness. This combination of properties may be obtained in a number of steels, when they have been subjected to the appropriate heat treatments. Steels containing more than 0.6 per cent of carbon are termed tool steels (see section 9.7 and table 9.1). When a high-carbon steel is quenched in water or oil from some temperature above the lower critical point the structure formed will consist of cementite and martensite. The material in this condition will be extremely hard, but also very brittle. Tempering the quench-hardened material at some temperature within the range 200–300°C will increase the toughness of the material, but this will be at the expense of some of the hardness. Typical tempering temperatures for plain-carbon tool steels are given in table 13.1.

TABLE 13.1

Tempering temperatures for plain-carbon tool steels

Tempering temperature (°C)	Applications
220	Hacksaw blades
230	Planing and slotting tools, hammers
240	Milling cutters, drills, reamers
250	Taps, shear blades, punches, dies
260	Stone-cutting tools, drills for wood
270	Axes, press tools
280	Cold chisels, wood chisels, plane blades
290	Screw drivers
300	Wood saws, springs

For many single-point cutting-tools it is customary only to heat treat the tool tip. This is accomplished by rapidly heating the tool tip in a flame until it is red hot. There will be some temperature rise in the shank material during this stage of the treatment. The tool tip is then rapidly quenched. After quenching there is heat flow from the shank and this is sufficient to bring the tip to the required temper.

In this way the tool cutting-edge is hardened and tempered while the shank remains in a soft and very tough condition.

One major disadvantage of plain-carbon tool steels is the comparatively low temperature at which tempering and softening will occur. Tool temperatures in the range of 250-300°C are easily reached in turning and milling operations at comparatively low cutting-speeds.

The addition of alloying elements to steels will cause alterations in properties. Both tungsten and chromium form very hard and stable carbides. When added to steels both elements cause an increase in the critical temperature values and also both elements bring about an increase in tempering, hence softening, temperatures. High-carbon steels rich in these two elements can be hardened and tempered, but they retain their high hardness even when heated up to temperatures of 600-700°C. This makes the materials very suitable for use as cutting tools capable of high cutting-speeds. Consequently, they are termed *high-speed tool steels*. A tool made from this type of steel will retain its edge and continue cutting even when the tool tip has reached dull red heat (about 600-650°C). The best known type of high-speed tool steel, termed 18/4/1 steel, contains 18 per cent tungsten, 4 per cent chromium and 1 per cent vanadium, with a carbon content of about 0.8 per cent. The heat treatment for high-speed tool steels differs from that for plain-carbon steels and involves the following stages

(1) soak at 800°C
(2) heat to 1300°C and air quench to harden
(3) temper at 600-650°C for desired hardness

The extremely high hardness and the retention of hardness at high temperatures of metallic carbides and oxides would make these materials very suitable for cutting purposes. These ceramic materials are, however, non-ductile and brittle, but they can be used as cutting tools when backed by a strong tough material. Tungsten carbide was the first ceramic material to be used in the manufacture of tipped cutting-tools. The tungsten-carbide particles are cemented together by means of an iron–cobalt alloy. This type of material, known as hard metal, was originally made by hot pressing tungsten-carbide powder together with cobalt and iron. Subsequent developments have given hard metals composed of a mixture of carbides, such as tungsten carbide, titanium carbide and tantalum carbide, in a matrix of cobalt. Cobalt is an exceptionally good binder for the carbides since it forms a thin continuous film around all the carbide particles. With cobalt contents of up to 15 per cent the material is wholly elastic and brittle. Alumina (aluminium oxide) has also been used as a cutting-tool material.

13.5 Shearing and Blanking

Shearing operations using a guillotine-type machine are very often used to cut sheet, plate, rod or wire to the required dimensions. Blanking is a similar type of operation but in this case the blanking tool cuts, by shearing, a designed shape from flat sheet or strip material, for example the production of circles from sheet metal for subsequent deep drawing or pressing. If the shape that is cut out from the sheet is scrap material, as in the production of slotted angle, the process is called piercing, although it is identical with a blanking operation.

There is a certain clearance between the stationary and moving blades in a guillotine, and between punch and die in a blanking press. When the sharp edges of the tools contact the metal during shearing there will initially be some bending and this will lead to the generation of cracks on both surfaces of the sheet. The cracks will develop and meet and, in so doing, produce a clean cut (see figure 13.3).

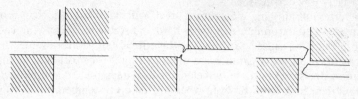

FIGURE 13.3 *Shearing*

The extent of tool clearance is very important in shearing and blanking operations. If the clearance is too small two large jagged edges may be formed on the metal. Not all materials behave in the same manner. Soft, ductile materials will plastically deform to a considerable extent before fracture begins and a fairly large tool-clearance is necessary for shearing and blanking, whereas hard materials tend to crack with little or no prior plastic deformation and tool clearances should be small. Tool clearances tend to be between 5 and 12 per cent of the thickness of metal being cut.

13.6 Grinding

Grinding is in effect a metal-cutting operation with a multipoint tool. The grinding wheel is composed of small hard abrasive particles cemented together with each hard particle acting as a cutting tool. Because of the small size of each cutting particle in a grinding wheel, each swarf chip produced is exceptionally small and continuous chips are not formed. Soft materials give rise to difficulties, since small particles may flow and fill in the spaces between the abrasive constituents of the grinding wheel. This would quickly clog the wheel and render it largely ineffective for grinding. Hard brittle materials, on the other hand, may be readily ground.

The principal applications of grinding as a machining process are

(1) for shaping extremely hard metals
(2) for shaping when very little material needs to be removed
(3) for producing a good and accurate surface finish
(4) for machining ceramics

Non-cutting machining processes, such as electrolytic machining and spark machining, may be used for shaping hard and brittle metals but they are not applicable to ceramic materials.

14

Joining Processes

14.1 Introduction

There are very many instances in which materials have to be joined to one another. In very many cases joining is part of the fabrication process for a complete article. The shape of the finished component may be too complex to manufacture in one piece and several small parts may have to be assembled and rigidly joined. Alternatively a component may be too large to make in one piece and again several parts may have to be joined together. The joining processes that may be used in assemblies include the use of bolts, screws and rivets, and also bonding by means of soldering, brazing, welding and the use of adhesives and cements. Only bonding processes will be discussed in this chapter.

Joints between metals may be made using the processes known as *soldering, brazing* and *welding.* In the two former processes the bonding agent is a metallic alloy that is different from the metals being joined, while in welding a filler material, if used, is of the same or similar composition as the metals which are being joined. Some thermoplastic materials may be welded, and many plastics materials, together with wood and ceramics may be bonded using adhesives and cements.

14.2 Soldering and Brazing

Soldering and brazing involve the use of a bonding, or filler, alloy that possesses a lower melting-temperature than the melting temperature of the metals to be joined. The solder or brazing metal must be capable of wetting and alloying with the metals to be bonded.

Soft solders, with melting temperatures in the range 150–300°C, are basically alloys of lead and tin but they may contain some antimony. The surfaces to be soldered must be clean and oxide-free otherwise the liquid solder will not properly wet and alloy with surface material. This means that in addition to preparing the surfaces by cleaning and degreasing, the use of a soldering flux is also necessary. The purpose of the flux is to act as a solvent for surface oxide-films. Zinc-chloride solution or resin fluxes may be used for soft soldering. Lead–tin solders will bond with steels and copper but to aid the soldering of steel it is often useful to pre-tin the steel surface.

Soft solders based on lead and tin cannot be used in connection with aluminium. An aluminium solder, which is a eutectic alloy of aluminium and silicon with a melting temperature of 560°C, may be used for jointing in aluminium.

Brazed joints are made using brass, a copper–zinc alloy, as the bonding medium. Brazing is similar to soft soldering but the brazing alloy melts over the range 860–900°C, and the resultant joint is considerably stronger than a soft-soldered joint. Alloys of copper, zinc and silver are also used for jointing purposes. These alloys, known as the *hard solders*, or *silver solders*, are considerably stronger than brass and also melt at lower temperatures. The silver solders melt in the general temperature range 700–850°C. An oxide solvent flux is needed in connection with brazing and hard soldering, in the same way as for soft soldering, and the most widely used brazing flux is borax.

Furnace brazing is a method often used in the manufacture of components on a mass-production scale. Parts to be joined are assembled with a thin strip or disc of brass placed in the joint area. The assembled components are then placed in a controlled-atmosphere furnace and heated to a temperature of 950–1000°C. At the high temperature the brass melts and effects the joint. The use of a flux is not necessary since the furnace atmosphere is oxygen-free.

During soldering and brazing operations the parts being joined do not become molten. These processes are used when low- or medium-strength joints are required and it is not desirable to use very high temperatures which could result in changes in structure or shape distortion.

14.3 Fusion-welding Processes

In fusion welding some portion of the metals to be joined are molten during the process and the filler metal is of the same or similar composition as the metals being joined.

The combustion of acetylene in oxygen at a specially designed burner is capable of producing extremely high temperatures. The oxy-acetylene torch has long been used as a heat source for fusion welding. Other hydrocarbon gases can also be used as a welding heat-source and oxy-propane and oxy-butane torches are frequently used. The relative proportions of oxygen and gas at the burner nozzle can be varied to give a flame that is chemically neutral, or oxidising, or reducing. A chemically neutral flame is used for the welding of most ferrous and non-ferrous metals, but a slightly oxidising flame is preferable when welding copper, brass and nickel silver to prevent the absorption of hydrogen in the weld. A neutral or slightly reducing flame is used for the welding of aluminium alloys to minimise oxidation of the aluminium. This also applies to some alloy steels. Oxy-acetylene welding is widely used, particularly for welding mild steel. The oxy-acetylene flame is also a very useful tool for cutting steels. A fierce oxidising flame is used and steel sheet or plate thicknesses of up to 30 mm may be cut through by a process of localised melting. For cutting through greater thicknesses a different design of burner is used permitting a much greater flow of oxygen. The steel is heated up to ignition temperature and then combustion of steel in the oxygen jet

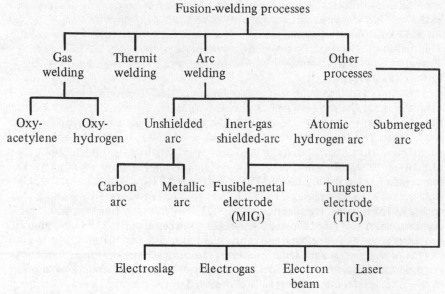

Fusion-welding processes

continues, cutting through the section. The force of the jet is sufficient to blow away molten steel and iron oxides. Thicknesses of up to 1 m may be cut in this manner.

Thermit welding is sometimes used for *in situ* welding of sections such as rails. A small mould is built around the parts to be joined. Thermit mixture (aluminium powder and powdered iron oxide) is contained in a reaction crucible placed above the mould. The powder is ignited and the following reaction occurs

$$8\,Al\,+\,3Fe_3O_4\,=\,9\,Fe\,+\,4\,Al_2O_3 \quad \Delta H\,=\,-414.5\ MJ/kmol\ aluminium$$

The molten iron runs into the mould. Selected steel scrap turnings are usually mixed with the thermit mixture to produce a steel of the same composition as the steel being joined.

A very common source of heat for welding is obtained by striking an arc between an electrode and the workpiece. The early forms of arc welding employed carbon electrodes. In this method the filler material had to be fed separately into the arc. The carbon arc is rarely used nowadays and the electrode is generally of the consumable type, melting into the work to provide the filler material. Uncoated electrodes may be used but it is more usual to use an electrode coated with flux. The electrode coating serves several purposes and is formulated to exercise the following functions

    (1) to provide an oxide solvent flux
    (2) to stabilise the arc
    (3) to produce reducing gases which shield the metal during transference from electrode tip to weld pool, and
    (4) to form a protective slag layer over the completed weld

In metallic-arc welding the current supply may be either a.c. or d.c. and if d.c. the electrode may be either the negative pole or the positive pole. For a d.c. system the greatest heat penetration into the workpiece occurs for a negative electrode and positive workpiece. This polarity is widely used for the welding of steels, but in the welding of aluminium, stainless steels and other metals with tenacious surface oxide-films the welding electrode is made the positive pole in the circuit. This makes the electrons flow from the workpiece to the electrode in the arc. The rapid flow of electrons in this direction helps to fragment and disperse the surface oxide layer.

Most automatic arc-welding processes use uncoated electrodes with some form of shielding for the arc. In inert-gas shielded-arc welding a stream of inert gas such as argon is blown around the electrode. The inert gas forms a blanket around the weld area, preventing oxidation and allowing good welds to be made without a flux. The shielding gases generally used are argon or helium, although pure CO_2 is often used for shielding purposes in the welding of mild and low-alloy steels. The gas-shielding processes are termed MIG welding and TIG welding. In MIG welding the electrode is a bare metal wire. This acts as a consumable electrode and is the source of filler metal. The electrode in the TIG process is of tungsten and is non-consumable. A separate filler rod then becomes necessary. Both the MIG and TIG processes are suitable for hand use or in fully automatic installations.

One of the earliest shielded-arc processes to be developed was the submerged-arc process. This is suitable for the automatic welding of long straight weld-runs in the horizontal mode. Powdered flux is placed in the prepared joint. The tip of the electrode lies below the flux surface so that the arc is not exposed to air. During welding the flux melts and forms a protective slag layer over the weld.

The electroslag and electrogas processes are automatic processes for the welding of thick sections. In the electroslag process (suitable for sectional thicknesses of 50 mm up to 600 mm) consumable electrode wire is melted under flux cover by the heat generated in the passage of current through a conducting flux. The electrogas process is similar in some respects but instead of a conducting flux, welding is carried out in an atmosphere of carbon dioxide, and the process is suitable for sectional thicknesses of 8 mm up to 50 mm.

Electron-beam welding can only be carried out *in vacuo*. A high-velocity electron beam is focused on the area to be welded. Extremely high temperatures are produced locally. Very fine welds with small heat-affected zones can be produced, and also great depths of penetration can be achieved. No fluxes are needed since the process is *in vacuo*. An electron beam is rapidly dissipated in air, but the concentrated energy beam from a laser source does not suffer in this way, and the use of a laser beam as a heat source for welding is possible.

14.4 Structures of Weld and Heat-affected Zones

A fusion weld is, in effect, a small casting and its structure may vary considerably from that of the parent metal. During welding, portions of the work adjacent to the weld become heated to a sufficient extent to cause alterations in the structure, and hence the properties of the material. The areas in which property alterations occur are called the heat-affected zones.

Let us consider briefly the structural and property alterations that may occur in some typical types of material.

Example   A non-ferrous material such as aluminium which is in the cold worked and annealed condition before welding. The weld pool itself will solidify to give a cast structure that will probably possess a coarser grain structure than the parent metal. Portions of the parent metal on either side of the weld will become heated to such a temperature that some grain growth can occur, but the width of the heat-affected zone will probably be small. Since physical strength and hardness are related to grain size with an increase in grain size causing a reduction in strength, the actual weld and the heat-affected zones will be less strong than the parent material. This is shown in figure 14.1a.

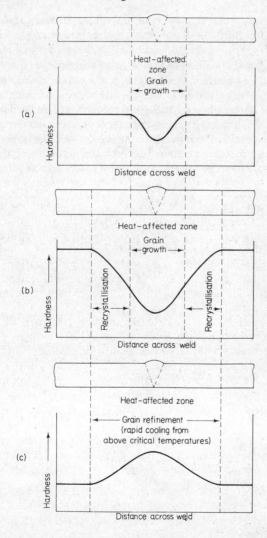

FIGURE 14.1 *Structural and property changes across butt welds:* (a) *for an annealed non-ferrous metal,* (b) *for a cold-worked non-ferrous metal,* (c) *for annealed or normalised mild steel*

Example   For a similar material to that above, but a material that is in the cold-worked condition before welding, the pattern will be similar, but with one significant difference. The heat-affected zones will consist of all material adjacent to the weld that has been heated to the recrystallisation temperature of the metal. Within these zones the cold-worked crystals will have recrystallised giving a major reduction in hardness, and close to the weld there will be grain growth as in the first example. The total extent of the heat-affected zone will be greater than in the first case. This is shown diagrammatically in figure 14.1b.

Example   In steels other effects can occur in the heat-affected zones, so let us consider an annealed or normalised mild steel as the third example. (Refer to the iron–carbon phase diagram figures 9.11 and 9.12). The weld metal begins to solidify as δ ferrite, but this will quickly change to austenite. The actual weld will probably cool through the critical temperature range quite rapidly causing austenite to transform into a fine-grained ferrite and pearlite. This zone, with a smaller grain size than the original annealed or normalised structure, will be harder and stronger than the parent metal. The heat-affected zones will be those portions that are heated above the lower critical temperature during the welding operation. The material within these zones will be converted into austenite during the welding operation and will transform on cooling into fine grained ferrite and pearlite (see figure 14.1c).

With higher-carbon steels and some alloy steels the rapid cooling of weld metal could cause some brittle-martensite formation. One of the functions of a welding flux is to provide an insulating layer over the completed weld, reducing the cooling rate and minimising the formation of martensite in these steels.

With some alloy materials there is the possibility of precipitation effects occurring in heat-affected zones and such precipitation could lead to trouble. The precipitated particles could cause embrittlement or a reduction in corrosion resistance, as is the case with some stainless steels.

During the welding of steels some oxidation of carbon occurs. To counteract this, the filler rod or consumable-electrode rod should have a slightly higher carbon-content than the steel being welded. In the welding of medium- or high-carbon steels the generation of carbon dioxide could lead to porosity in the weld unless very great care is taken. Other defects that may occur in welds are incomplete fusion and entrapment of oxide and slag particles in the weld. Non-destructive testing techniques such as ultrasonic testing and radiography are frequently used to check the quality of welds.

Thermal stresses may be established during the cooling and contraction of some welded structures. In severe cases the residual stresses could be high enough to cause post-weld cracking. Where, due to the type of material and weld design, this is possible, post-weld stress-relieving heat treatments have to be made.

14.5 Pressure-welding Processes

The oldest form of welding is *forge welding*, and this is suitable for both wrought iron and carbon steels. The parts to be joined are heated to bright red heat (about 1000°C) and then hammered together. The surface oxide-scale is broken up and metal crystals grow across the original interface. Wrought iron is self-fluxing, since the slag content acts as an oxide solvent flux, but for the forge welding of steels silica sand is used as a flux.

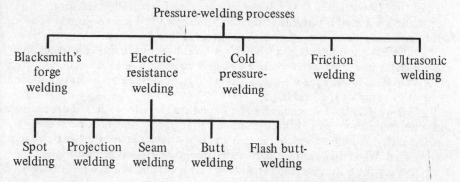

Electric-resistance welding in one form or another is very widely used as a production tool. In all the various methods the metal parts to be joined are made part of a low-voltage electrical circuit and the two components pressed together. Pressure is applied before a current is passed, and is maintained during the period of current flow and for a short period afterwards. The heat developed during the passage of the electric current is sufficient to render the metal locally plastic and to allow crystal grains to grow across the interface making a good joint. The pressures used in these welding processes are in the range 75–400 MN/m² and current densities of up to 80 A/mm² are necessary. The length of time required to make a spot weld is usually less than 0.5 second. Steels are very easy to weld by these methods but high-conductivity metals, such as copper and aluminium, are more difficult.

In *spot welding* the parts to be joined are clamped between a pair of water-cooled copper electrodes (see figure 14.2), current is passed for the necessary time and pressure maintained for a further short period. Spot-welding machines are usually equipped with accurate timing controls that can be pre-set to the desired values.

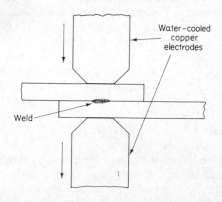

FIGURE 14.2 *Spot welding*

For much factory production multiple spot-welding machines are used in which from 2 to 100 welds can be made simultaneously. This is the production method used for the manufacture of such things as car-body assemblies and pressed-steel hot-water radiators. While spot welding is generally used for joining light-gauge sections it is possible to use the technique for thicker sections. Spot welds have been made in steel sections of up to 25 mm thickness.

Seam welding is similar in prinicple to spot welding, but in this case overlapping sheet metal is passed between rotating wheel electrodes. A series of welding-current pulses is passed through the work producing a continuous seam which is in effect a series of overlapping spot-welds.

In *butt welding* the two surfaces to be joined are butted together under pressure and a welding current passed through the material (see figure 14.3). *Flash butt-welding* differs from butt welding in that an arc is struck between the two surfaces to be joined. When the heating effect of the arcing current has been sufficient to begin melting the butt ends the current is switched off and the two hot surfaces are brought rapidly together under pressure.

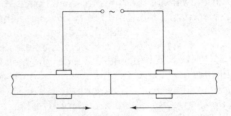

FIGURE 14.3   *Butt welding*

Cold pressure-welding between surfaces can occur if the surfaces are pressed firmly together and also sheared so that there is some sliding between the two surfaces. This type of pressure welding is not greatly used but it can be suitable for the joining of some thin foils.

In the process of *friction welding* one of the parts to be joined is placed in a chuck and rotated. The other part is placed in a fixed chuck and is pressed firmly against the rotating member. Considerable frictional heat is generated at the rubbing surfaces and seizure and welding takes place. This is a method capable of producing very good welds and has the advantage that it can be used for joining dissimilar metals. Ultrasonic vibrations can be used as a means of causing one surface to slide against another and so weld together. This method is particularly suitable for welding thin foils, and again may be used for welding dissimilar metals.

14.6 Welding of Plastics

A number of thermoplastic materials may be joined by welding but the thermo-setting plastic materials are not weldable once they have been fully cured. Fusion welding is possible for any material that melts without charring over a regular and fairly well-defined range of temperature. Because of the low thermal conductivities of thermoplastic materials they are difficult to heat effectively by external means but some materials, including polyethylene and rigid PVC, may be fusion welded using a hot-gas torch as the source of heat. This type of torch is a heat exchanger. A stream of neutral gas, which may be air or nitrogen, is heated in the body of the torch and emerges from the nozzle as a hot gas-jet. A filler rod is melted into the joint, in a similar manner to the use of a filler in the welding of metals, but no flux is required.

Analogous methods to electric-resistance, spot- and seam-welding techniques for metals are based on dielectric heating. Thermoplastics are electrical insulators and when a thermoplastic is placed between two capacitor plates and the plates connected to a high-frequency source there will be a heating effect within the dielectric. With an intense field strength and frequencies within the radio range the heating effect is sufficient to cause local melting of thin material. Spot welds can be made using circular capacitor-plates. If rotating discs are used as capacitor plates then continuous seam-welds can be made. This type of welding is used for PVC sheet material.

A number of thermoplastic materials may be joined by means of *solvent welding*. A thin film of volatile solvent is spread over both surfaces to be joined. After the solvent has begun to dissolve both surfaces, the two parts are pressed firmly together. This allows the solutions to mix together thoroughly. When the solvent evaporates there is an excellent bond between the two original parts.

14.7 Use of Adhesives

An adhesive, glue or cement is a liquid or paste that will thoroughly wet the surfaces to be joined and then harden. When a liquid wets the surface of a solid there will be some bonding across the interface. This bonding is probably of the van der Waals type and when the van der Waals forces are strong, due to the presence of highly polarised molecules, the liquid may form the basis for a useful adhesive material. For good adhesive bonding it is also useful if the solid surfaces are not perfectly plane but contain crevices or other surface roughness for good keying. In the glueing of wood, for example, there is penetration of glue into the wood cells. When plywood is case hardened the outer edges of the wood cells are bent over and closed up. It is almost impossible to make a good glued joint on this smooth type of surface.

Glues may be solids of low melting point that are melted, applied and allowed to solidify, or solids in solution that harden after application by evaporation of the solvent, or cold-setting resins in which the resin and hardener liquids are mixed just before application and hardening (curing) then takes place slowly. The rate of hardening may be increased by applying heat.

Carpenter's bone glue is a gelatine made from animal or fish bones and has a melting point of about 80°C. This type of glue is applied hot and quickly cooled and set. It possesses a number of drawbacks. The gelatine is water soluble and hence is not suitable for external use. It can also support fungal growth. Nowadays most adhesives are synthetic high-polymer materials. The phenol-formaldehyde- and urea-formaldehyde-type thermosetting materials have been used but their use is largely confined to the manufacture of plywood and other laminates because they require heating to about 150°C for curing.

One of the best range of solvent-type adhesives is based on the linear polymer polyvinyl acetate and these materials find wide application for the joining of timber, paper, card and other materials.

The most widely used adhesives are the cold-setting polymers, in which adhesive and hardener are mixed together immediately prior to application. These materials are based on either polyester resins or epoxide resins. On full curing the material forms a full network rigid polymer structure of high strength. Adhesives of this type can be used for joining most types of material including metals, ceramics, glass, wood, leather and some types of thermosetting plastic .

15

Materials Selection

15.1 The Selection Problem

The question of what material is to be used for the manufacture of some particular component is often an extremely complex problem. There are very many factors which have to be considered, including

(1) the shape and size of the component
(2) the dimensional tolerances to be attained
(3) the scale of production required
(4) the physical and mechanical property requirements
(5) cost and availability of materials
(6) cost and availability of processing plant

These various factors cannot be viewed in isolation from one another since there is a complex interaction between them. There are of course some instances where the range of materials suitable for a particular application is limited. For example, if the requirement is for a material with an electrical resistivity of less than 2.0×10^{-8} Ωm the only available materials are high-conductivity copper and silver. The latter metal is normally ruled out because of its high cost, and copper becomes the solution. If the conductivity requirement is less stringent, say a conductivity $\not> 3.0 \times 10^{-8}$ Ωm is allowable, then another material, high-purity aluminium, could provide a viable choice. Even then the selection is straight-forward, being largely a question of balancing the lower cost of aluminium against the greater efficiency of copper as a conductor. Selection problems are seldom as straightforward as this.

There is a major interrelation between materials selection, production processing and design. The choice of material and processing route should be made at an early stage in the design study. This means that the production technologist, the designer and the materials expert must be involved as a team. The type of material selected will influence the choice of processing route and vice versa. If the material choice is mild steel, production by diecasting is ruled out. Conversely, if die-casting is the chosen manufacturing route the materials available are restricted to the lower-melting-point metals and alloys. Decisions on both manufacturing route and choice of material will affect many detailed aspects of the design. Also the size and general shape of a design will obviously have an influence on the choice of processing route, and hence on the choice of material.

The mechanical properties of a material may provide a guide to the suitability of the material for some particular service environment, but they are not always an accurate reflection of behaviour in service. As an example, nylon 6.6 possesses a low hardness value compared with many metals, yet small nylon gear-wheels are capable of performing satisfactorily in many situations where, formerly, metals of high hardness were specified.

15.2 Forms of Supply

Metallic ores are smelted to produce metals. The purity of the metal produced is dependent on the nature of the metal, the quality of the ore and the type of smelting process employed. For example, the smelting of iron ores in a blast furnace produces a pig iron that may only contain from 85 to 90 per cent of iron, the balance comprising carbon, silicon, manganese, sulphur and phosphorus. The refining of iron is part of the steelmaking process and the impurity content of blast-furnace iron is considerably reduced in the conversion to steel. The aluminium smelter, on the other hand, produces aluminium with purities of 99.5 to 99.8 per cent. This product may be refined, if necessary, to produce aluminium of 99.99 per cent purity.

'Pure' metals, of varying purities, are generally available in ingot form. Many primary manufacturers also carry out alloying and are able to supply the casting industry with ingots conforming to various alloy-specifications in addition to pure-metal ingots.

Wrought metals are available in many forms including sheet, strip and plate in many widths and gauges, bar, tube, wire and extruded sections in many patterns. A major user of wrought metal will be able to purchase direct from the manufacturer. In many cases the manufacturer will produce and supply material to the exact requirements of a bulk consumer and, of course, there tend to be special price arrangements for such orders. The customer whose requirement for, say, metal sheet is small generally has to purchase his supplies from a metal stockist who, in turn, obtains his supplies in bulk from a primary producer. The metal stockist will tend to concentrate on specifications and sizes that are in fairly constant demand and it is often difficult for the small customer to obtain supplies of those alloys for which general demand is low.

Plastics materials are available in the form of powdered or granulated moulding compound. The primary plastics producers compound the raw plastic with the necessary fillers, plasticisers and dyes and the moulding compounds are then in suitable form for the manufacture of injection or compression mouldings, extrusions, etc. Thermoplastic materials can also be obtained in the form of sheet, film, rod and tubing.

The firm that produces articles for sale does not necessarily design and manufacture all the components or sub-assemblies that go to make up the finished article. For example, when a new car is designed certain items such as lamps or wiper motors, may simply be bought 'off the shelf' from another producer. Similarly, a wiper motor may be designed to incorporate bushes and gears that are available 'off the shelf' from yet another firm. The various trade directories contain details of the manufacturers and suppliers of the many and varied stock items such as rivets, screws, bushes, gears, etc., that are used in the manufacture of sub-assemblies and finished products.

15.3 Cost Aspects and Availability

The ex-works cost of a finished component, or product, is compounded of many ingredients, including the cost of raw material bought in, labour costs and fixed overheads for plant and machinery, etc. It is impossible to generalise but in many manufacturing industries the purchase cost of raw materials may account for 50 per cent of the works cost of a finished component. Generally, one could expect the selection of a cheaper raw material to give a lower-cost final product. This is often the case, but in some cases the reverse may be true. A more expensive raw material may require fewer processing operations and the final works cost may be less than when using a cheaper raw material.

Raw-material cost is often expressed per unit mass of material. This is not always the best means for drawing comparisons between one material and another. In a large number of instances the general dimensions of a component may be determined by design factors and the mass of the component may be relatively unimportant. Comparisons based on cost per unit volume then become of greater value than cost per unit mass. Many thermoplastics are much more expensive than mild steel on a cost-per-unit-mass basis, but not when considered on the basis of cost per unit volume. It is sometimes valid to draw cost comparisons between materials on other bases. Material-cost-per-unit-property data is often helpful when attempting to solve a materials selection problem.

Every time that a material is subjected to some processing, cost will be added, and the costs of raw materials bear very little relation to the costs of semi-finished products such as cold-rolled steel sheet. Similarly, the costs of an alloy are generally higher than the cost of the principal basis metal in the alloy. The increases in cost for some steel and aluminium semi-finished products are given in tables 15.1 and 15.2. The figures quoted in these tables are those that applied in mid-1975. While the costs of raw materials are constantly changing the principle that work performed creates a cost increase does not change, and so the cost of cold-rolled mild steel strip will always be higher than the cost of hot-rolled mild steel strip.

The availability of a material is also a major factor when arriving at a materials-selection decision. It is no use choosing some particular material for an application, even if it fully satisfies every design requirement, if sufficient supplies are not readily available. This question of availability also applies to manufacturing facilities. A firm will be loath to implement a design based on the injection moulding of a plastic, as an alternative to one based on machining components from metal bar stock if the former choice means investing in new moulding equipment or sub-contracting the moulding to another firm, while at the same time leaving its own large and well-equipped machine shop and a skilled labour-force idle.

TABLE 15.1

Cost build-up (steel products) (based on prices ruling in June 1975)

Material		Cost £ per tonne
iron from blast furnace		71
mild steel (ingot)		84
mild steel (black bar)		130
mild steel (cold-drawn bright bar)		175
mild steel (hot-rolled sections)		120
mild steel (hot-rolled strip)		123
mild steel (cold-rolled strip)		128
alloy steels		
black bar	1% Ni steel	186
	3% Ni; 1% Cr steel	234
	$3\frac{1}{2}$% Ni; 1% Cr; 0.3% Mo steel	255
bright bar	1% Ni steel	248
	3% Ni; 1% Cr steel	300
	$3\frac{1}{2}$% Ni; 1% Cr; 0.3% Mo steel	325

TABLE 15.2

Cost build-up (aluminium products) (based on prices ruling in June 1975)

Material	Cost p/kg
aluminium 99.5% ingot	39.6
aluminium commercial-purity sheet	54.5
plate	59.5
circles (for deep drawing)	66
simple extruded sections	75
cold-drawn tubing	86
aluminium alloy N4 (2% Mg) sheet	59
cold-drawn tubing	110
aluminium alloy H15 ($4\frac{1}{2}$% Cu) sheet – W condition	85

The preceding paragraphs serve to indicate that the factors influencing the selection of a material for an application are many and complex, and that frequently some factors may oppose one another. Often there is no perfect solution to a design and selection problem. In many instances there may be several possible solutions of almost equal viability. It is the author's hope that readers will accept the necessity for considering possible materials and process routes from the very commencement of any design study.

15.4 Case Studies

1. Figure 12.3 shows the individual castings and the complete assembly of a saw-grinding attachment. All the parts are precision die-castings in a zinc alloy. The knobs of the adjusting screws are cast direct onto the screws. The whole of this unit was specifically designed on the basis of production by die casting in a zinc alloy, but the cost of producing a component to fulfil the same purpose using other processes involving machining operations would probably be about twice the cost of producing this design.

2. Despite the fact that the cost of metal powders per unit mass is considerably higher than the cost of ingot metal the production of parts by powder-metallurgy processes can frequently lead to the unit cost of components being about 50 per cent less than the cost of producing parts by rolling or forging followed by machining operations. Some parts produced by powder techniques are shown in figure 12.14.

3. There are several suitable materials for the production of high-strength forgings for aircraft undercarriage components. A comparison of a high-strength low-alloy steel against a high-nickel-content maraging steel shows that the raw material cost of the latter is almost five times as great as that for the low-alloy steel. Yet the final works cost of a component made from a maraging steel may be between 5 and 15 per cent less than that for a low alloy steel because of the very much smaller machining costs involved.

FIGURE 15.1 *Combined gear–cam for an office calculating-machine, showing the original component made from five metal parts and the new component which is single moulding in nylon 6.6. (Courtesy of I.C.I. Plastics Division and Bell Punch Co. Ltd)*

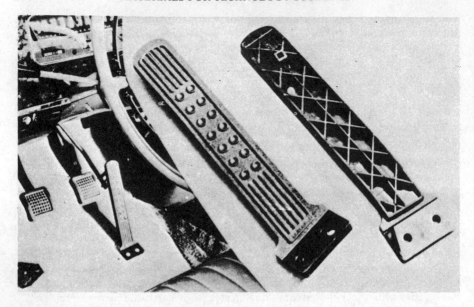

FIGURE 15.2 *Polypropylene accelerator pedal with integrally moulded hinge. (Courtesy of I.C.I. Plastics Division and the Plastics Division of Hills Precision Die Castings Ltd)*

4. Figures 15.1 and 15.2 illustrate how complex assemblies may be replaced by a single injection-moulding. Figure 15.1 shows a combined gear–cam, formerly made by assembling five separate metal parts, made as a one-piece moulding in nylon 6.6. Figure 15.2 shows a car accelerator-pedal made as a one-piece moulding in polypropylene. This design makes full use of a peculiar property of polypropylene, namely its ability when in thin section to be bent repeatedly without fatigue failure occurring. This allows the pedal hinge to be part of the one moulding. In both of these examples considerable cost savings are effected.

5. Design changes may frequently be progressive. A pulley for use in a washing machine was formerly made as an iron casting, which needed a considerable amount of machining. The part was redesigned as an aluminium diecasting, but some machining was still required. This change resulted in a reduction of 4.7 p in the cost per component. A further design change, making the pully wheel as a two-piece diecasting in a zinc alloy, requiring assembly of the two parts, but no machining, gave a further saving in works cost of 4.1 p.

APPENDIX A

Macro- and Microscopical Examination of Metals

A.1 Macroexamination

Macroexamination — examination with the naked eye or at a low magnification — is a useful technique for the inspection of fracture surfaces, and for determining some of the characteristics of metal structures.

Much useful information can be obtained from visual examination of a fracture surface with the naked eye. The difference between a tough fibrous fracture and a brittle cleavage-fracture can be readily observed. A fatigue fracture also possesses a characteristic appearance. It may also be possible to detect the presence of slag inclusions and porosity on a fracture surface, and such defects may have been the points of initiation of the failure. This type of surface detail will be revealed with greater clarity if the surface is further examined with the aid of a low-power (up to × 50 magnification) stereo microscope.

In order to obtain information about the structure of a metal it is necessary to section the material. The cut surface is then ground flat and etched with a chemical reagent. A high degree of surface finish on the specimen is not essential for macroexamination, and grinding can be finished with a grade 0 emery paper. During the etching treatment the surface layer of metal is removed and there is a preferential attack on certain constituents and inclusions. There are very many macro-etching agents and the choice of agent will be dependent on the nature of the metal and the type of feature that it is desired to reveal. Details of some macro-etchants are given in table A.1. Macroexamination after etching will reveal such detail as defects, segregation effects, and fibre structure in wrought metals.

Sulphur printing is a macroexamination technique that is suitable for the examination of plain-carbon steels. Steels contain small sulphide inclusions, and the distribution of these inclusions within the steel is a guide to the distribution of all non-metallic inclusions. The sample of steel to be examined is sectioned and the cut surface is ground flat. A piece of bromide photographic-paper is soaked in a 3 per cent solution of sulphuric acid for two minutes and it is then carefully placed, with the emulsion side downward, on the prepared steel surface. The sulphuric acid reacts with sulphide inclusions forming hydrogen sulphide gas. This gas reacts with the silver bromide in the emulsion forming a brown deposit of silver sulphide. The bromide paper should be in contact with the steel surface for about three minutes. The bromide paper is then removed from the steel, washed in water, and the print is 'fixed' by immersion in a 'hypo' solution for a few minutes. Dark brown areas on the print indicate areas containing sulphide inclusions in the steel section.

TABLE A.1

Some etchants for macroexamination

Composition	Method of application	Uses
Hydrochloric acid 140 ml sulphuric acid 3 ml water 50 ml	Immersion in solution for 15 to 30 min at 90°C	A deep etch for steels
Copper ammonium chloride 9 g water 91 ml	Immersion in solution $\frac{1}{2}$ to 4 hours	To reveal dendritic structures in steels
Ferric chloride 25 g hydrochloric acid 25 ml water 100 ml	Immersion	For copper and its alloys
Ammonium persulphate 10 g water 100 ml ammonium hydroxide 50 ml	Immersion	For copper and its alloys
Hydrofluoric acid 20 ml water 80 ml	Swab surface of sample with etchant	For aluminium and its alloys

A.2 Microexamination

Microscopic examination of metals is used to reveal fine details of structure. Metals may be examined at magnifications of up to × 2000 with the aid of an optical microscope. In order to examine the structure of a metal at high magnification it is necessary that the metal specimen be carefully prepared, and because metals are opaque to light, incident illumination must be used. Oblique incident illumination is perfectly satisfactory for macroexamination, but normal incident illumination must be used for examination at magnifications of × 50 and greater. The construction of a metallurgical microscope differs from that of a biological microscope (using transmitted light) to cater for this. The metallurgical microscope possesses a built-in light source. A partially reflecting mirror is situated in the microscope tube, and this will reflect the illuminating light through the objective lens on to the specimen surface (see figure A.1). Alternatively a small reflecting-prism may be used in the microscope tube.

A metal specimen for microexamination is prepared by cutting a small but representative sample from a metal component, followed by grinding and polishing a surface of the specimen to a mirror finish. This is achieved by a series of successive hand grinding-operations using progressively finer grades of paper. The specimen should be rotated through 90° at each change of paper. The final grinding should be on grade 500 or 600 silicon carbide paper. Paraffin should be used as a lubricant for the grinding of very soft metals. After grinding is complete the very fine surface scratches are removed by polishing the surface to a mirror finish. The polishing powders that may be used are alumina, jeweller's rouge or magnesia. These polishing powders are suitable for the polishing of steels and

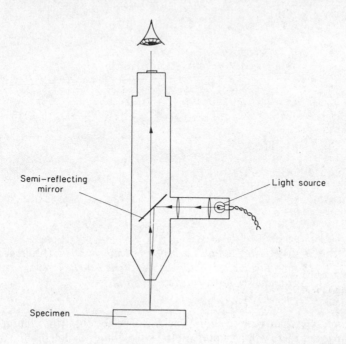

FIGURE A.1 *Principle of metallurgical microscope for normal incident illumination*

many other materials, and are used in water suspension in conjunction with a Selvyt or broad-cloth polishing cloth. The polishing cloth is mounted on a flat rotating disc and the specimen is held against this under light pressure. The proprietary metal polishes 'Brasso' and 'Silvo' are highly suitable for polishing copper and copper alloy specimens. Pastes containing very fine diamond dust (particle sizes ranging from $\frac{1}{2}$ to 3 μm) are also widely used for metal polishing. Alternatively, the finely ground surface may be polished electrolytically. This entails anodic solution of the ground surface in a suitable electrolyte.

The flat polished surface of the specimen should be examined under the microscope. The highly polished metal surface will appear bright and will show no structure, since it acts as a mirror to the normal incident light, but the following features, if present, can be observed

(1) cracks and porosity
(2) non-metallic inclusions (slag inclusions, and graphite in cast irons)
(3) hard constituents, that stand out in relief from the matrix

To reveal the complete structure of the specimen it is necessary to etch the polished surface in a dilute chemical reagent. The choice of etchant will depend on the nature of the material and the type of feature being investigated. Details

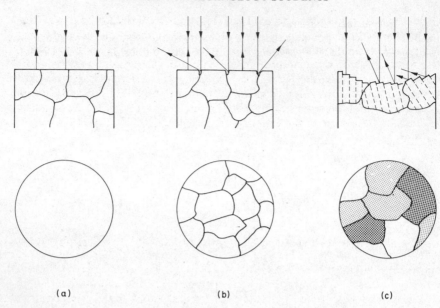

FIGURE A.2 (a) *Polished metal surface, no detail revealed;* (b) *etched surface revealing grain boundaries;* (c) *etched surface revealing grain boundaries, but with grain shading*

of some micro-etchants are given in table A.2. The action of the etching agent will be selective and there will be a preferential attack on crystal grain-boundaries and at constituent boundaries. Examination of the etched specimen under the microscope will now show the grain boundaries revealed (see figure A.2). The effect of an etching agent is not usually a general solution of the metal surface.

TABLE A.2
Some etchants for microexamination

Etchant	Uses
Nital: 2 ml nitric acid in 100 ml industrial alcohol	Excellent general etchants for irons and steels, other than stainless steels
10 g ammonium persulphate 20 ml ammonium hydroxide 80 ml water	Good general etchant for copper and its alloys
10 g ferric chloride 30 ml hydrochloric acid 100 ml water	Good etchant for copper and its alloys; provides more contrast than ammoniacal ammonium persulphate solution
5 ml hydrofluoric acid 1 ml nitric acid 100 ml water	Etchant for aluminium alloys; surface of the specimen should be swabbed with cotton wool soaked in etchant
1 g sodium hydroxide 100 ml water	General etchant for aluminium alloys

Often the attack produces a faceted surface, and the orientation of the surface facets may vary from grain to grain. In this case the amount of reflected light recaptured by the objective lens of the microscope will vary from one crystal grain to another. This will lead to some crystals appearing to be light in colour and other crystals appearing dark, even though all the crystals may be identical in composition and type. Alloy constituents in multi-phase alloys show up clearly after etching and the form and distribution of phases can yield much information about alloy properties to the experienced observer.

A.3 *The Electron Microscope*

The wavelength of visible light is of the order of 5×10^{-5} m and this means that it is impossible to resolve objects of smaller size than about 1 μm using an optical microscope. For magnifications in excess of about $\times$ 2000 it is necessary to use the electron microscope. Electrons possess wave characteristics and the wavelength of electrons in a 100 kV electron microscope is about 3.5×10^{-12} m. In an electron microscope a beam of electrons is produced by an electron 'gun'. The electron beam passes through an evacuated tube and is focused on to the specimen by means of magnetic lenses. The specimen is inserted into the vacuum chamber via an air-lock. The electrons pass through the specimen and through the magnetic objective and projector ('eye-piece') lens systems of the microscope to be projected on a fluorescent screen. Magnifications of up to $\times$ 100 000 may be obtained with electron microscopes, and objects of sizes of the order of 10^{-9} m may be resolved. Because the electron microscope uses a transmission technique and metal specimens readily absorb electrons, metal samples can only be examined directly if in the form of very thin foils (foil thicknesses of the order of 10^{-6} m). These thin foils of metals have to be prepared very carefully. This is usually done by thinning a small piece of the metal by chemical or electrolytic solution. Transmission electron-microscopy of metal foils has been used as a research technique since the mid-1950s. Before that date metal samples were not viewed directly and replica techniques were used. A replica of an etched metal surface can be obtained by depositing a thin film of a plastic on the surface. The replica can then be stripped from the metal surface and viewed in an electron microscope. Both replica and thin-foil techniques are used today.

Another type of electron microscope that is used as a research tool is the scanning electron-microscope. This type of microscope can accommodate a fairly large specimen and is used for examining the surface topography of a material, for example, a fracture surface. A very fine electron-beam scans the surface of the specimen, in a similar manner to the scanning of a television picture. The electron beam is scattered from the specimen surface and scattered electrons are picked up by a collector. The signal produced from the collected electrons is used to modulate the scanning beam of a cathode-ray tube and this produces a picture of the surface area of the specimen under examination. One major advantage of the scanning electron-microscope is that it possesses a large depth of focus. Magnifications of up to $\times$ 40 000 are possible with this type of instrument.

APPENDIX B

Data Tables

TABLE B.1.

Physical properties of some pure metals

Metal	Symbol	Melting point (°C)	Density (kg/m³ × 10⁻³)	Crystal structure	Elastic constants (GN/m²) E	G	Poisson's ratio	Transformation temperatures (°C)
aluminium	Al	660	2.7	f.c.c.	70.5	27.0	0.34	
antimony	Sb	630	6.67	r.				
beryllium	Be	1280	1.85	c.p.h.	313	160	0.28	
bismuth	Bi	271	9.8	r.	32		0.33	
chromium	Cr	1888	7.1	b.c.c.	238	88.5	0.3	
cobalt	Co	1492	8.7	c.p.h.	203	75	0.31	
				f.c.c.				430⁺
copper	Cu	1083	8.9	f.c.c.	122.5	45.6	0.35	
gold	Au	1063	19.3	f.c.c.	80	28.3	0.42	
iron	Fe	1535	7.87	b.c.c.	215	84.8	0.29	
				f.c.c.				908
				b.c.c.				1388
lead	Pb	327	11.3	f.c.c.	16.5	5.6	0.44	
magnesium	Mg	649	1.74	c.p.h.	44	17.6	0.28	
manganese	Mn	1244	7.4	c.cub.	200	77.6	0.24	
				c.cub.				700⁺
				f.c.t.				1100⁺
				b.c.c.				1140⁺
molybdenum	Mo	2620	10.2	b.c.c.	338	120	0.3	
nickel	Ni	1453	8.9	f.c.c.	208	78.7	0.31	
niobium	Nb	2420	8.57	b.c.c.	104	36.7	0.39	
platinum	Pt	1769	21.65	f.c.c.	173	61.2	0.39	
silicon	Si	1412	2.34	d.				
silver	Ag	961	10.5	f.c.c.	79	29	0.37	
tin	Sn	232	7.3	d.	40.8	20.6	0.33	
				b.c.t.				18
titanium	Ti	1660	4.51	c.p.h.	106.4	40	0.33	
				b.c.c.				880⁺
tungsten	W	3380	19.3	b.c.c.	393	152	0.36	
uranium	U	1130	19.05	orth.				
				t.				668
				b.c.c.				774
vanadium	V	1920	6.15	b.c.c.	127	46.8	0.37	
zinc	Zn	419	7.14	c.p.h.	92	37.3	0.25	
zirconium	Zr	1860	6.4	c.p.h.	95.5	36.4	0.33	
				b.c.c.				

Key: b.c.c. – body-centred cubic   b.c.t. – body-centred tetragonal
       c.cub. – complex cubic   c.p.h. – close-packed hexagonal
       d. – diamond   f.c.c. – face-centred cubic
       f.c.t. – face-centred tetragonal   orth. – orthorhombic
       r. – rhombohedral   t. – tetragonal   + = approximate value

TABLE B.2.

Comparison of properties of metals, plastics and ceramics

Property	Metals	Ceramics	Plastics
Density ($kg/m^3 \times 10^{-3}$)	from 2 to 16 (average 8)	from 2 to 17 (average 5)	$1-2$
Melting points	low to high (Sn $232°C$) (W $3400°C$)	high, up to $4000°C$	low
Hardness	medium	high	low
Machinability	good	poor	good
Tensile strength (MN/m^2)	up to 2500	up to 400	up to 120
Compressive strength (MN/m^2)	up to 2500	up to 5000	up to 350
Young's modulus (GN/m^2)	$40-400$	$150-450$	$0.7-3.5$
High-temperature creep resistance	poor	excellent	–
Thermal expansion	medium to high	low to medium	very high
Thermal conductivity	medium to high	medium but often decreases rapidly with temperature	very low
Thermal-shock resistance	good	generally poor	–
Electrical properties	conductors	insulators	insulators
Chemical resistance	low to medium	excellent	generally good
Oxidation resistance at high temperatures	poor except for rare metals	oxides excellent SiC and Si_3N_4 good	–

Index